The
ABUNDANT
PEACE

The
ABUNDANT
PEACE

Robert C. Garretson

The World Publishing Company
CLEVELAND AND NEW YORK

Published by The World Publishing Company
2231 West 110th Street, Cleveland 2, Ohio

Published simultaneously in Canada by
Nelson, Foster & Scott Ltd.

Library of Congress Catalog Card Number: 64–8565

Second Printing

Dedication

To PRESIDENT JOHN FITZGERALD KENNEDY

—who hoped that the United States would be respected around the world for its civilization and not for its might, and who believed that when nations work together to improve their economies and the living standards of their people they are on the road to peace.

Author's Preface

NOT so many years ago—in 1933 to be exact—I was finishing my master's thesis, "Propaganda in Soviet Russia," at Northwestern University. I had returned in 1931 from a five-month assignment in Soviet Russia for Food Machinery Corporation where I had an opportunity to assist in installing machinery in two canning factories, a completely new one in Krimskaya and an older one in Krasnodar. I had learned to get along in the Russian language and had argued about Soviet plans for the future with many young Russian workers. During the course of that trip, I visited Leningrad, spent a week or more in Moscow, and had a short vacation in Anapa and Novorossisk on the Black Sea. On my return, my interest in Soviet propaganda led me finally to do a detailed study of every item published in the Moscow *Daily News* for an entire year. The *News*, published in English in Moscow, was an official Soviet publication.

Since those years I have drifted far away from the subjects of propaganda and Soviet Russia. Although I have found time to read about Russia and to keep informed of Soviet economic progress, it was only in 1961 that I decided to review more carefully the record of the last thirty-five years and to attempt a projection of current trends to the end of the twentieth century.

Being a romantic by nature, I thought of looking at the town of Krimskaya, a village of 15,000 where I had lived in 1930, and making a comparison of its growth and development with a similar town in the U.S.A. For purposes of comparison, I decided to use the somewhat larger city of Kalamazoo, partly because it had a name that began with *K*, and partly because I had been acquainted with Kalamazoo as long ago as 1929.

I began to gather all the statistical data available, and eventually decided that a trip back to Krimskaya would be essential to understanding what might have happened there in the last thirty-five years. *Stanitsa Krimskaya*, meaning a little village in the country, is now *Krimsk*, a healthy adolescent in the Krasnodar area. It now has a population of 30,000 and is a real town, which is why the suffix *aya* is no longer used. It now has an important *combinat*, or collection of local industries, under the leadership of Leonid Semonov, a vigorous young man whom I met in 1963.

Krimskaya—Krimsk—is in the southern part of Russia, near the Black Sea, fifty miles or so from the seaport of Novorossisk. It is near the delta of the Kuban River, that flows into the Sea of Azov, and is very near the fabulous Crimea of Yalta and Sevastopol. Only a few miles south the trailing ends of the Caucasus mountains separate the northern plain from the vineyards along the Black Sea. Krasnodar, on the Kuban River, is a city of about 400,000 population in 1964.

My visit in 1963 took me first to Moscow, where I had prearranged interviews with specialists for the canning and brewing industries and with economists attached to the institutes for economic research. I also had a good chance to explore

Moscow, from the University to the permanent Exhibition on the other side of town, and from Archangelskoe to Abramtseva, two fabulous old estates that are now converted to public museums. I saw the ballet twice, at the old Bolshoi Theater and at the new Kremlin Congress Theater—Moscow's Lincoln Center.

From Moscow I went to Kharkov and from there to Krasnodar and Krimsk. My return trip was via Odessa and Kiev to Vienna. All along the way I had the good fortune to find people who were helpful in giving me a better idea of the changes that have been made since 1929.

Despite my curiosity concerning Krimsk and Krasnodar, and my interest in the achievements of the Soviet Union in the last thirty-five years, my reasons for writing this book are far from romantic. What I want to know is what the prospects are for further Soviet growth, and what the economic trends in the U.S.S.R. may mean to us over here. This is why I have examined nearly all the available statistics, why I went back to visit Moscow and Krasnodar and Krimsk last fall, and why I have put down on paper this record of what I was able to find out and what it means to me.

The information that I have obtained concerning the U.S.S.R. has come almost entirely from the excellent studies published by American economists who have been trying for a number of years to rationalize the statistics available concerning Soviet Russia, so that they could be compared with similar figures for the United States. In addition, I did have a chance to discuss my projections with several Soviet economists. All of these gentlemen were very friendly and cooperative, and my thanks go to them for their help even though they did not agree in all cases with the figures that I have used. In particular I would like to mention:

Mr. Andrei Izotov, Food Industries Department, U.S.S.R. Economic Council. Mr. Izotov is a chief specialist for the canning industry in Moscow. He was born in 1901, spent a year or two in the United States in 1930-31, and returned to Stanitsa Krimskaya, where he was chief engineer for the new canning factory.

Mr. V. F. Krushinin, a chief specialist for the brewing industry in the U.S.S.R. Economic Council.

Vice Director Ivan Mayevsky of the Economic Institute of the Academy of Science in Moscow.

Mr. Boris Urlanis, also of the Economic Institute of the Academy of Science, who has written a book on the dynamic structure of population in the U.S.S.R. and the U.S.A.

Mr. Joseph Budnitsky, also of the Economic Institute of the Academy of Science, whose specialty is capital investment and economic mining production.

Mr. G. M. Popov, from the section of information for the Institute of the Academy of Science.

Professor Anitoli Shapiro, with the Institute of World Economics and International Relations of the Academy of Science. Mr. Shapiro was especially interested in the projection of current statistics into the future.

My comparisons of the Soviet and American economies are based on information that was obtained from relatively few sources. A key comparison is that of the gross national products (GNP's) of the two countries in 1964. A figure of $250 billion in 1954 dollars for the U.S.S.R. provides the base from which other figures can be deduced or projected, using the percentages in Abram Bergson's *The Real National Income of Soviet Russia Since 1928* (Cambridge: Harvard University Press, 1961) *and the Comparisons of the United States and Soviet Economies* (86th Congress, 1st Session, Joint Economic Committee, Parts I, II, and III).

The $250 billion figure is rounded and approximate, but is a reasonable estimate in the view of Soviet as well as of American economists.

Other valuable sources used were:

Schwartz, Harry, *Russia's Soviet Economy*, 2nd ed. (Englewood Cliffs, N. J.: Prentice-Hall, 1954).

Campbell, Robert W., *Soviet Economic Power* (Boston: Houghton Mifflin, 1960).

Spulber, Nicolas, *The Soviet Economy: Principles, Problems* (New York: W. W. Norton, 1962).

Rostow, W. W., *The Stages of Economic Growth: A Non-Communist Manifesto* (London and New York: Cambridge University Press, 1960).

Wilcox, Weatherford, and Hunter, *Economies of the World Today: Their Organization, Development, Performance* (New York: Harcourt, Brace & World, 1962).

The projections I have made backward and forward from the available figures are my own responsibility, but they are based on conservative estimates of the current trends. I have chosen to take a rather longer look into the future than most economists would consider to be scientific because I feel strongly that a short look fails to reveal the full meaning of the trends that are discernible in the U.S.A. and in Soviet Russia today. What could possibly happen in thirty-five years may be more meaningful and significant than a more accurate estimate of what will happen in the next ten years.

Nearly all of the figures I have used for the economy of the United States of America are taken direct from the published series of the National Industrial Conference Board, although in some cases I have gone directly to Government sources. I am also indebted to Mr. Samuel L. Bennett of the W. E. Upjohn Institute for Employment Research for statistical information concerning the growth of Kalamazoo in the last thirty-five years. In using 1954 dollars as a constant measure of economic growth and change, I am following the practice of the N.I.C.B. rather than attempting a second conversion to 1964 dollars. Where I have used current dollars instead of constant dollars, this has been indicated in the text. The projections made into the future are conservatively literal and mathematical. The figures have been pushed ahead at the same rates that have been achieved in the last thirty-five years, or in the case of Soviet Russia at more recent rates that are somewhat lower and appear to be more reasonable. All the correlations between Soviet and U.S.A. statistics have been made by projecting relationships that have been found in these sources. The objective has not been to verify any preconceived theories but rather to find out something about the prospects for these two major economies.

Many of the trends seem to be converging. Among these are per capita incomes, living standards, and industrial techniques. However, this does not mean that the attitudes and points of view of the two economies are necessarily converging. Nor is it certain that the present trends will continue as projected. Political and cultural pressures must be reckoned with, and there are many alternative courses that either economy could take. All I am saying is that if my premises turn out to be correct, this is what could happen.

Some bias enters into all calculations, but I have been as unbiased as possible in reaching my conclusion that the economic trends in the U.S.A. and the U.S.S.R. lead in the direction of an abundant peace.

Contents

II. TRENDS PROJECTED TO 1999
U.S.A. AND U.S.S.R.

The
ABUNDANT
PEACE

Introduction

I F it were possible to look ahead thirty-five years to 1999
and see what the economies of the U.S.A. and the U.S.S.R.
would be like, it might be possible to avoid a thermonuclear
war. It would be possible to make better economic plans and
to save wasteful competition that didn't work or wasn't necessary. It would also be possible to do without many threats,
showdowns, murders, and sacrifices that proved nothing at
all in the end.

Although it is impossible to look ahead thirty-five years,
or five years, and see things as they certainly will be, it's not
impossible to look back thirty-five years and find the trends
that are leading into the future. It is possible to ascertain
what has happened, to note the changes that have occurred,
and to sort out the significant areas of growth and development that might be expected to continue in their present directions.

The objective of this study is to look at the facts, to measure them as accurately as possible, and to find the economic trends that may be projected ahead to determine where the economies of the U.S.A. and the U.S.S.R. are going, and what kind of problems they will have to face and solve in the future. In operating a business it is not unusual to project current trends ahead five to ten years, and although businessmen know that these projections must be checked and corrected regularly, they also find that in many cases their original projections of five or ten years ago were surprisingly close to the mark. They have learned to respect the facts, and to use all of the statistical information available, but they have learned as well how to cast the dice, and to assess the chances of some things happening that can't be measured. They have learned to gamble on mathematical probability and not on sheer luck or personal desires and opinions. They have thus learned now, with the trends in hand, to make intelligent plans for a profitable future.

To extend this technique to the historical future of civilizations that appear at this time to be engaged in hopeless competition is not as reckless or incredible as it may seem to be. It is not much different from Toynbee's technique of looking backward fifty or a hundred years following a trend he thought he had discovered to see if he could find the reasons and causes that theoretically should have been there to explain what happened much later. Happily, he could generally find something to verify or deny the projections he had made. It won't be possible to verify similar projections into the future except as they can be measured year by year against what actually does happen. If it should turn out in some cases that what happens does agree with the guesses that have been made, then those projections will gain in credibility and usefulness. If large annual adjustments must be made in other projections, these will be viewed with increasing skepticism.

Even Toynbee, with fifty years of study, could know only

a fraction of what there was to know of the world history that he studied. Factors multiply, and there can be no end of knowledge.

The idea is what matters, the hypothesis, the well-conceived, partly logical, and partly intuitive guess that can be checked and reviewed and used for more guesses, new ideas, and a different understanding of man, time, and the universe.

Seen in ultimate perspective, knowledge itself is no more than the current guess, verified by its context, and its framework of experience and practice. The areas of experience and practice change continuously and knowledge today is not what it was yesterday or what it will be tomorrow because as its frame of reference changes, it changes. Even logic and mathematics change, as equality and identity are better differentiated, and as other than numerical values are added to old equations. $A = B$; $B = C$; therefore, $A = C$ isn't necessarily so. And $2 + 2$ may be more than 4 with the time dimension reckoned, and the value of time counted.

Consequently, it turns out that our dreams of putting all the knowledge acquired through the ages into a computer so that we can then have the advantage of using the mechanical brain to solve our problems with full reference to everything that is known about any subject can never be realized in practice. Only current knowledge can have meaning in solving current problems. The knowledge accumulated through the ages turns out often enough not to be knowledge today. It has changed and become myth and is either not true or not relevant to current situations. No matter how much knowledge might be gathered and stored in a machine, no use or meaning could be made of it without relating it to an idea or a hypothesis, and old knowledge could have little relevance to new hypotheses.

One reason for looking ahead to see where the trends are going is to find out whether they are taking the world where it wants to go. If present trends continued for another hundred years, the total population of the world would exceed four and one-half billion, and total world production would be at least ten times as large as it was in 1964! The question is whether there would still be starvation in India and China, and whether the average person in the U.S.S.R. would have more than fifteen square meters of living space. Not to look ahead to see what might happen would be both stupid and reckless. Before the pressures of fear and frustration precipitate total nuclear destruction, it would be wise to examine the current state of the economy and to estimate what the situation may be even before the present century is over.

Thirty-five years ago it would have been impossible to guess what the U.S.A. and the U.S.S.R. would accomplish in only three short decades. As I remember, nobody did guess, not even the Russians. In 1964 the situation is quite changed. Both the U.S.A. and the U.S.S.R. have acquired formidable banks of statistics and both have learned to use incredibly ingenious electronic computers. Even a relatively practical businessman can make projections that are safely within the field of probability, if he will use the most reliable information at hand and let the trends he finds in the past run to their logical conclusions in the future.

The question of absolute accuracy and consistency in the statistics used either to measure the past or project the future is not relevant to the arguments developed in this study. What matters is the direction of the trends, not the absolute rate of change. What matters is the attitudes displayed by the two economies—the concern demonstrated for economic growth, employment, and automation. It must be recognized from the start that all statistics are less than perfect. Numbers simply can't measure all the changes that there are from one year to the next. Quality changes, functions change, and needs change. The same total of goods and services won't have the same value or usefulness next year that it has this year and surely the same number of people have different skills, different desires, and different economic meaning from one decade to the next.

Yet, the figures are important in indicating the directions of change and in estimating at any one moment the relative value attached to different sections of the economy.

For example, if asparagus accounts for 5 percent of the sales of vegetables in this year compared with 1 percent thirty years ago, it is fair to say that asparagus has gained in favor as a marketable vegetable. If sales in dollars have increased ten times in the last thirty years, this does not mean that ten times as much asparagus is being consumed or that ten times as many people are buying asparagus. The present asparagus may be bigger and better and more expensive, or it may be that many more people are living in the country and growing their own asparagus, or it may be that new processing and transportation methods are making asparagus available more continuously and more extensively to the same number of asparagus fanciers. Inevitably, a variety of factors will affect the validity of any deduction taken from figures for different years and under different circumstances.

If this is true within the U.S.A., it is even more true when it comes to comparing two entirely different economies like those of the U.S.A and the U.S.S.R. The figures are essential to getting an intelligent appreciation of what is happening. They should not be treated lightly but neither should they be taken literally.

It can be argued logically that the population of the U.S.A. will reach a figure of 325 million by the end of the century and that there will be 375 million in the U.S.S.R. However, it is not illogical to project somewhat lower figures of 300 and 350. In my judgment, when there are nearly as good reasons for alternative sets of figures, it is best to make a choice based on the factors of believability, intelligibility, and communicability, as well as probability. Certainly the figures least likely to favor the proposition being considered should be preferred. This is why conservative estimates of future growth have generally been selected for this study when there were reasonable alternatives.

The main question in the world today is whether the U.S.A. and the U.S.S.R. will be able to coexist successfully for the rest of this century and live happily together thereafter. The answer to this question is not just a matter of economics, but if it is possible to know what the economic situation is likely to be in the U.S.A. and the U.S.S.R. in 1999, it will be much easier to estimate the probable effect of other social and political factors.

The idea that the U.S.S.R. could dominate the world is based on the assumption that it could achieve economic leadership. Is this fear well founded or is it unrealistic? Is there any practical meaning in the concept of world economic domination? How different are the economies of the U.S.A. and the U.S.S.R.? Do the differences create additional reasons for mistrust and eventual conflict?

These are questions that need to be answered with facts instead of fears.

The facts available will be used to measure and compare the trends that have developed during the last thirty-five years in the U.S.A. and the U.S.S.R. Then these trends will be projected ahead another thirty-five years to get an approximate idea of what the economic situation may be in 1999. All of these projections will be made on the assumption that peace can prevail for this period of thirty-five years. In total, they will amount to an estimate of the economic gains that can be made in the U.S.A. and the U.S.S.R. if there is peace.

ONE

Economic Trends Since 1929—
U.S.A. and U.S.S.R.

I

Economic Perspective

A THOUSAND photographs of the U.S.A. and the U.S.S.R. as they look in 1964 might be interesting and instructive, but they would be of little value in measuring the different rates of growth or the direction of change guiding these economies. No series of facts or observations concerning the situation in the U.S.A. and the U.S.S.R. at a single moment of time could be very helpful or useful for appraising the meaning and the prospects for these economies that may be contained in the present moment. Without perspective the pictures are flat, and there is no life in the people, or in what they are doing. It is necessary to know where they are going and where they have been.

Perspective is another word for the fourth dimension. Time is a factor that is often overlooked entirely, or that is measured incompletely or inconsistently. Yet, time exerts powerful pressures on all of the other factors in every political and cultural, yes, and economic equation. The changes, the results, the new balances and relationships that must be expected in even a few years can greatly alter the conclusions that might be drawn from a picture of the economies of the U.S.A. and the U.S.S.R. exactly as they are today. The differences that might occur during a period as long as thirty-five or fifty or one hundred years could completely redefine the economic problems that seem to exist today, and might suggest solutions that can't even be imagined in 1964.

This is why it is necessary to look back thirty-five years in order to look ahead the same distance. It is necessary to get an idea of the direction of the changes that have led to the present situation, and to see the alternate courses that may be taken, and that can help to solve our pressing immediate challenges, and contribute to the longer-range solutions that are required of all of us, Russians and Americans alike, as part of a continuing human race.

A far-reaching perspective doesn't matter very much to a lunatic, a religious fanatic, or an opportunistic commissar or stockbroker; nor to a madman like Hitler or an egomaniac like Mussolini. Perspective doesn't matter to a man trapped in a cave with a tiger, or to men anywhere ready to die to escape infamy or tyranny. Men motivated by emotion aren't concerned with perspective or logic or truth. Such men may still destroy the world.

But an accurate perspective does matter to those who care to see and understand *our* problems, and for the moment let's think of us as all the people who will inherit the earth in the perspective of the next hundred years. The economic trends in the U.S.A. and the U.S.S.R., running back from 1964 to 1929, amount to a perspective that will help to clear away some of the haze that partially obscures the prospects for the future.

Let's look back a few years to 1929. Is it possible that there were millions hungry, if not starving, in the richest country in the world?

Does anyone remember Rush Street on the near north side of Chicago where the poets and intellectuals gathered to discuss socialism and art? To the north and east was the Gold Coast, where few could see what was happening. To the south and west was poverty, and even starvation.

It was the same in Cincinnati and Baltimore and Boston. Only thirty-five years ago a quick depression and a violent one proved how thin and unstable the economy of the U.S.A. really was. It wasn't a long depression, only a few years, but it didn't hit bottom until 1933. Few people actually died of hunger. In Chicago there was even stale bread to be had to feed riding horses for less than the price of oats—bread that the A & P couldn't sell. Just that long ago there was no insurance against unemployment nor any social security nor any plan at all for the national economy!

The automobile had nearly replaced the horse and buggy, and the first assembly lines were turning out additional cars at a rate of one every five minutes or so. In most factories machines were doing a fair share of the work, and in a few industries the possibilities of completely automated production were seen vaguely in the future.

Men worked five and one-half days every week in most shops and factories, perhaps forty-five hours, and they were glad to work as many extra hours as possible with no thought of a higher rate of pay for overtime. In Wisconsin women were permitted by law to work only nine hours a day in canning factories where they were paid 22¢ per hour. They earned at most $1.98 per day, which would buy about as much as $4 in 1962. A good cook and maid lived in for $10 per week!

Before 1929 the economy had moved ahead steadily for fifty years. This had been a period of expansion across the country and although the industrial revolution had made significant changes in the way of living of most people, the economy was still heavily weighted toward agriculture.

The population of the U.S.A. nearly doubled from 1890 to 1929, increasing from 63 million to 122 million. During the same time, the gross national product (GNP) was up 384 percent, or 200 percent per capita, from $750 to $1,492 in 1929 (all in constant 1954 dollars). This increase of 200 percent in thirty-nine years compares with an increase of 180 percent in the thirty-five years from 1929 to 1964.

If 1933, the low point of the depression when per capita GNP fell to $1,007, were used as a starting point, the increase from 1933 to 1964 would be 266 percent and the increase from 1890 to 1933 would be only 134 percent.

The rate of growth before 1929 had been tied for a hundred years to the growing population, with a continuous shifting of jobs and chores from the kitchen, sewing room, and barn to industry. It is impossible to measure how much of the increase in the gross national product is not an increase in fact, but a shift from work done at home or on the farm, which could not be valued in dollars for time and materials, to work done in shops and factories. The actual increase in living standards was not as great as the increase of per capita GNP would suggest.

The men who ran this economy were practical men, tradesmen and farmers, many of whom had never gone to college. They were the self-made men who had learned the business from the ground up. Often they had worked through the factories after the fashion of apprentices, and had taken their places as managers upon reaching an age of experience if not sobriety. Most managers were mature men, not only in age but also in practical experience, and many of them were part owners of their enterprises. Not only was a college degree not necessary to get a start in business in 1929, but it was only just beginning to be a help. Big business was already big, but its share of the total economy was still relatively small, and even in very large corporations, most often decisions were made on the basis of experience and intuition, without recourse to statistical analysis and market research. This was a practical economy, which, in 1929, found itself unable to understand or to cope with the economic depression that turned it upside down without warning.

Meanwhile, in Soviet Russia the first five-year plan was getting under way. For the first time an economic plan for an entire nation had been formulated, and the people of the nation were being assigned to the specific jobs that had to be done. Whether they liked it or not, the Russian workers were going to build an industrial economy. They were going to invest their labor in new factories and machines even though they might do without shoes and enough to eat. In fact, there was a famine, and many did starve, but the work went on just the same. There was no cutting back on the program for industrial expansion, and success was in sight when the war with Germany upset all of the Soviet calculations.

In 1929 many workers in the U.S.S.R. did not own a pair of shoes. It's hard to believe, but in Krimskaya where Food Machinery Corporation was putting machinery in a new canning factory, a number of the young Russian workers had to tie rags around their feet or go barefoot. Many did get shoes or boots that summer, but the supply of shoes in the single local store was never adequate to satisfy the needs, and it was always necessary to get an order from the village authorities before a purchase of items in short supply could be made. I will never forget the satisfaction I had in getting a pair of boots for one of my workers who had nothing to put on his feet. Luckily, we both wore about the same size.

Nineteen hundred twenty-nine was the twelfth anniversary of the October Revolution of 1917. Soviet Russia was twelve years old, an age reached by the U.S.A. before 1800. It was a time of dedicated devotion by the young people and the party members, and of cynicism and apathy by many of the older people, a time of hope and doubt when no one knew with any certainty what would happen next.

Total Soviet production of $65 billion was about equal to that of the U.S.A. in 1900.

The record of growth in the U.S.A. from 1929 to 1964 reflects a new economic philosophy that was first tentatively accepted in the thirties. What had happened was that a period of steady and consistent economic growth had come to an end after nearly one hundred years. In the depression of 1929 to 1933 per capita production in the U.S.A. dropped below the level achieved in 1910, a setback of twenty-five years. For the first time government intervention in the affairs of business became necessary, and for the first time, the government joined business leaders in thinking about plans and controls for the national economy. The depression brought about what was almost a revolution in the attitudes of economists and businessmen toward the operation of an industrial economy as increasingly complex as that of the U.S.A. had become.

Reluctantly, it was agreed that the ideas and ideals of laissez-faire and "let the buyer beware" were inadequate for an industrialized economy. It was recognized that means must be found to employ or provide for all of those who could and would work, and to ensure a market for all of the goods that could be produced. Price controls, unemployment insurance, social security benefits, economic codes of behavior for industry, and experiments in regulating supply and demand were tried with varying degrees of success. What was tried, and how it worked is not as important as the fact that new results were achieved by trial and error, and by testing economic theories in practice.

It is not too much to say that a new kind of industrial economy got under way in the U.S.A. in the 1930's. The economy that began then might even be called a new phase of the industrial revolution. It became an economy in which the government accepted many responsibilities that previously had been left entirely to private business, or to the immutable and inviolate laws of economics, the economics of free trade. The idea of tampering with these economic laws was as shocking in the thirties as the idea of international free trade would be today. Nevertheless, the experiment was begun, and the economy began to move again at a new and exciting rate of growth.

It may be argued that the progress of the last thirty-five years was not the result of new attitudes toward economic theories, and would have come about anyway because of the development of new machines and new techniques, and that the pressure of a World War was more important than anything else in accelerating these developments. An even more cynical point of view is that the production of armaments required for the war, and continued since the war, was the major reason for the exceptional rate of growth since 1933. There is some merit in both of these arguments, but they are beside the point.

During the thirties the swing from farm to factory was well on the way to being completed. The manufacture of nearly all products had moved from the home or small shop to the larger units of specialized industries. There were a few dressmakers and milliners left producing finished products from raw materials, but the shoemaker had become a shoe *repair* man, and even the blacksmith only fitted and shaped the horseshoes he was able to buy already made in various weights and sizes. Agriculture had become adjusted to large-scale mechanized operations, and was moving rapidly in the direction of the specialized efficiency that was becoming a symbol of the new industry.

The time had come for economists, industrial leaders, and managers to turn their attention to improving and perfecting the industry that had been created, and to finding and solving the economic problems that the new industrial enterprises were facing. Their first objective was to cut costs and save labor by mechanizing every possible step in their manufacturing operations. The trend toward complete automation thus began as long ago as 1933. Next, the need for some controls on the economy to limit unemployment and to resist inflation became apparent after the dramatic stock-market crash of 1929. The trend toward more government control and responsibility got under way. It was soon evident, too, that careful planning for economic growth based on a thorough analysis of all of the economic factors would have to replace to a large degree the daring and opportunism of the preceding pioneer years.

The obvious first step to take was to make sure that every product was produced as efficiently as it could possibly be produced with the aid of all of the mechanical and scientific knowledge available. During most of the last thirty-five years, business has been concerned primarily with production problems. It has given most of its time and research budgets to improving the efficiency of production techniques, processes, machines, and automatic controls. The goal of industry has been to produce good products in tremendous quantities at the lowest possible costs. Only secondarily, and more recently, has industry become seriously interested in the problems involved in marketing the goods that are produced. It is important to remember that before 1929 the problem of selling what was manufactured was relatively easy to solve (hire an extra man—or buy some advertising). Up to that time the main objective had been to supply basic needs that were far in excess of the available supplies. The economy remained a frontier economy. From 1929 to 1964 attention was turned to production, even as the ability to produce more goods at lower prices was realized, and as, with only a little prodding, markets grew fast enough to absorb all that could be produced.

It is now a long time since 1929. Following the frightening setback of 1929 to 1933 the economy has moved forward at an unprecedented rate. Continued economic prosperity is reflected in all of the trends that can be measured statistically.

However, the rate of progress in the U.S.A. cannot match the record of the U.S.S.R. In 1964, the estimated per capita GNP in Soviet Russia of $1,096 is actually ahead of that of the U.S.A. in 1933, an increase of 258 percent over the per capita GNP of $425 in 1929, and the greatest part of this growth has been accomplished since 1943 in a period of only twenty years.

The U.S.S.R. now boasts that its total industrial production will exceed that of the United States of America in another *twenty* years. The fact that Russia's production is compared with that of the United States suggests that it must be catching up with the United States. This is hardly true, because it hasn't been behind the United States. It hasn't even been in the same race. It is important to remember in evaluating Russia's economic success that the Soviet Union hasn't simply overtaken other industrialized nations by getting over the same ground more quickly. It has in fact followed an entirely different path than anyone else. It has jumped across time by moving almost at one step from an economy of serfdom which never existed in the United States to a new kind of industrialized socialism that would be equally out of place in the United States. The U.S.S.R. has moved right over the familiar kind of capitalistic industrialism to something original and unique that could come about only in a command economy where the available supply of technically trained workers would be put to work with the newest techniques of industrial production, without regard for capital costs, wage scales, and standards of living. The new techniques required highly skilled technicians, and masses of semi-skilled workers. The U.S.S.R. set out to educate and train the workers it needed, even though it meant doing without adequate food and housing. It needed scientists and teachers. They were produced without regard to cost. It needed new factories and new machines and tools. All of its resources were concentrated on getting whatever was required, including bureaucratic management and political advisors. Its industrial production has increased at a nearly unbelievable rate, and whether or not Soviet Russia overtakes the United States in another twenty years, it is clear that it has succeeded in creating an industrial economy potentially as large or larger than that of the United States of America.

That the social, political, and cultural structure of a nation or civilization is largely a result of the economic situation in that nation may not be true, although it is a plausible hypothesis. It is impossible even to know for sure which is the chicken and which are the eggs. It is demonstrable, however, that there must be some connection between the kind and quality and number of production units in a society and the political and social institutions of that society. The spread of socialism in modern industrial states must be connected in some way with the volume of goods produced, the need for more customers, and the impossibility of permitting complete poverty and starvation to exist in a society that is able to produce more than enough goods to give everyone a passable living.

It may be that the potential failure of communism is connected in a similar way to the need for managers, skilled scientists, and other specialists in a modern industrial economy.

Whatever the reason is, it would seem that Soviet Russia has cut straight across the curve of economic development in the world as a whole to land at a point to one side of and possibly beyond that of any previous economy. Here it may well stand for a while to build and develop a new kind of economy that will probably continue to have a marked socialistic look, but that may contain some incongruous aspects of both capitalism and imperialism. Whether the United States is moving along a curve in the same direction is hard to guess. It may be, and it may arrive at a very similar economic structure in a few more years. It seems more likely, however, that the United States is following a different curve! Although it will have to face the same economic pressures of automation and modern technical skills, it is moving from a different base, economically, historically, socially, and philosophically.

That the U.S.A. will continue to expand its program for social security, and will continue to adopt one plan after another for eliminating poverty, aiding education, and insuring all of the people against the hazards of illness and accidents seems certain. Whether or not this is a trend toward socialism, each year more and more benefits are being guaranteed to all Americans.

In any case, the U.S.S.R. is moving very rapidly toward a modern industrial economy that may well surpass that of the United States in total production of all kinds of goods in relatively few years. If Soviet Russia does continue to move in this direction, and does succeed in attaining its production goals, how will this success affect the Soviet political theories and practices? If the time comes when its total production is equal to twice that of the United States today, then every family in Russia will be able to have all the things an American family has now. This is difficult to imagine, but it could be true within not too many years. There could be available for the Russian people as many cars, refrigerators, TV sets, houses, patios, garages, lawns and lawnmowers, playrooms and bars, golf courses and yacht clubs as there are in the United States today. The total per capita quantity of such consumer goods could be equal to what is available each year in the United States. The goods might not and probably would not be the same things, and it is unlikely that they would be distributed among the people in just the same proportions. In the U.S.S.R., as in the United States, some families would have a larger share than others. Some would have suburban *dachas,* or country houses, and others would live in a few rooms in the cities. How the goods produced were distributed would provide a useful clue to the direction of the economy toward or away from communism. In an ideal communistic society the total goods would be divided more or less equally among the population, or "to each according to his need." In the kind of society that there is in the United States, whatever it may be, goods are supposedly distributed in proportion to what is earned, and to a great extent it is assumed that what is earned is in proportion to what is produced. In other words, each worker expects to get a fair share of the total goods, relative to what he is able to contribute to their production. How is it in Russia at present? What is the trend?

The present trends in Russia are interesting and instructive. The bulk of the new luxuries available in the U.S.S.R. are going to a very small percent of the people. Yet, all of the people are assured a minimum standard of living. There is no poverty and there is talk of free bread for all in the near future. It can be argued that most of the people above the minimum subsistence level and below the privileged class level receive a share of the balance of goods available pretty much in proportion to what they earn. There is still a wide spread between the living standards of the lowest income groups and those at the top, but if the trends since 1929 continue, this spread will continue to narrow.

Even the higher income levels have relatively modest living standards in the U.S.S.R. today; and these are the engineers, scientists, and politicians who are making major contributions to the growth of the national economy. It is possible that a ceiling will be imposed on these higher income groups, excluding the "royal family" of government officials, of course, so that eventually the lower incomes can be raised until they are nearly equal to those at the top. This would be possible and the trends in housing suggest that it is being considered. It seems impractical to anyone from the U.S.A. because it would mean that no one could have living standards like those of the top 20 percent in the U.S.A. even if Soviet production were twice as large as that of the United States in 1964. However, if everyone in the U.S.S.R. could be persuaded to settle for living standards within the range of the so-called middle two-fifths of the population in the U.S.A., it could be done.

If the present trends continue, all income levels will continue to share in the increased per capita production that is available for household consumption. The managers, directors, and technical specialists will continue to enjoy a somewhat higher standard of living than the ordinary workers, and it is likely that the size of this group will increase as industry continues to expand and the need for skilled technicians and managers grows. In 1929 there were still a few kulaks left and managers were clearly of a different class from the illiterate workers. There were few admitted class distinctions, but there were engineers, commissars, industrial workers, and peasants. In 1964 even these distinctions are less clear. There may still be an unequal distribution of the total available consumer goods for another thirty-five years or more, but it is likely that very few families, if any, will live as well as the families in the top income groups elsewhere in the world. The top income groups in the U.S.S.R. may live as well as the upper middle groups in the U.S.A., and the lower income groups as well as the lower middle class in the U.S.A.

Office workers, managers, and skilled laborers will have new cars and TV sets, and manual laborers and unskilled workmen will have older cars and older TV sets just as they do elsewhere today.

Even in 1964 there are no slums and tenements in the U.S.S.R. However, there are thousands of crowded, small apartments that will become substandard housing in another twenty years or less. There will surely be better apartments, some row houses or their equivalent, and an increased number of *dachas* in the suburbs for a very small percent of the population.

These trends of economic growth are not like any that can be found in the U.S.A. Different segments of the economy are moving ahead at different rates. Capital investment is going ahead regardless of cost, and consumer goods are produced in limited quantities and standard qualities and styles. Identical housing is allocated by square meters per capita. There are six brands of beer, graded by alcoholic strength, and two grades of canned peas, and education is free. There are no cheap or shoddy goods, and there are none as good as the best that can be found in Vienna or New York. Everything comes in a few grades—good, a little better, and not quite as good.

Even the similarities that can be noted in the distribution of income by occupations and types of work are probably misleading. As in the U.S.A., the farmer and agricultural worker is very near the bottom of the heap in total income, and the entrepreneur, manager, or skilled manipulator of economic units is near the top. The place of engineers and scientists may be higher than it was at a similar period in the United States, and electronics and automation will have a new and different impact on the kind of unskilled and semiskilled jobs to be done, as well as on the kind of technical education that will be required for a large share of the population. But these are not differences in kind; they are only differences in degree. They may have little effect, if any, on the long-range trends. They are speeding the pace of Soviet industrial development, but they will probably not greatly change the look of the industrial culture that will be achieved in the U.S.S.R. as these trends are followed to their conclusions.

There are, however, some factors that are different, and that may result in an entirely new kind of industrialized society in Soviet Russia. For one thing, there is no opportunity for anyone to get into the top income brackets through speculation. There is no stock market. There are no stocks or bonds or other opportunities for the investment of excess earnings. There is no way for the ordinary citizen to acquire a share in the industry he works for, or to speculate on its growth. Theoretically, he already owns a share in the total economy as a citizen of the U.S.S.R., and theoretically, he gets his share of its growth in his pay check. However, though speculation, as we are used to it, may be eliminated, it is unlikely that the man who would have been a speculator has been eliminated. It is likely that he will be found contriving somewhere and somehow to build up the segment of industry with which he is connected. He will probably get the rewards of successful speculation in a better home, a motor car and driver, and first choice of the vacation resorts on the Black Sea. Nevertheless, this is a difference that will change in some way the character of the group who will make up the higher income class. Other differences have to do with the degree to which a top income can be inherited, the almost complete subsidization of education, and the controlled emphasis on certain aspects of the culture. The drive to produce the finest athletes to represent the U.S.S.R. in international competition is a case in point. A similar emphasis could be placed on art or music or cooking. Control of the channels of communication can condition the minds of the Soviet people to accept different values than would be found in other similar materialistic, industrialized societies.

Which is to say that although present trends may and probably will continue, the effect of a similar level of per capita production in Soviet Russia may not be the same as it has been in the United States and in other capitalistic countries. It seems likely that Russia is veering away from her original communistic target, but it is not yet certain exactly where she is going. When she is as well off as the United States, which may possibly be in less than fifty years, if not in twenty as her leaders boast, her economy may be more like that of the United States than could have been guessed a few years ago.

Although it is impossible to get statistics that are exactly comparable for the economies of the U.S.A. and the U.S.S.R., it will be helpful to look at several aspects of these economies to find out what trends there are and how they differ. Some that are significant in the U.S.A. may not be important in the U.S.S.R., but an examination of the same trends will point up the differences, and will indicate the economic relationships that may be achieved in 1999.

2

Economic Growth

ALTHOUGH production in the U.S.A. has increased nearly threefold since 1929, it has not kept pace with the rate of growth in the U.S.S.R. There, production increased nearly 400 percent from an estimated $65 billion in 1929 to $250 billion in 1964. Production in the U.S.A. was about three times that of the U.S.S.R. in 1929, and in 1964 it is only twice as large.

The GNP per capita increased from $1,492 to $2,680 in the U.S.A. and from $425 to $1,096 in the U.S.S.R. In 1929 the U.S.A. per capita GNP was three and one-half times that of the U.S.S.R. and in 1964 it is less than two and one-half times as large.

These are statistics, the best that can be found, but it is hard to know what they mean. In Moscow some economists[1] argued that the increase in production in the U.S.S.R. has been much more than 400 percent. For some products they

[1] All of the basic statistics for the U.S.A. used in this section were taken from tables prepared by the N.I.C.B. or directly from publications of the Department of Commerce, Bureau of the Census. Some of the computations were

have figures to show it has been more than 2,000 percent. However the Soviet indices don't include services and can't be compared directly with the gross national product of the United States. In any case, differences in kind and quality can't be measured in figures alone. It was my impression after talking to several Soviet economic specialists that my estimates are very near the truth. The net increase in the economy of the U.S.S.R. may be greater than these figures show, but, if so, the poverty of the Russian people in 1929 was even more severe than I could have guessed from my observations in 1930. In fact, very few meaningful statistics for the 1930's are available.

What is most stimulating about talking to the economists in Moscow is the evident fact that they are knowledgeable, sensible people who know what you and they are talking about and who are interested in getting at the truth.

Although I think they are too optimistic about the future, I don't think they are unscientific in their calculations. It is my guess that they are not allowing sufficiently for the lags and failures that will occur for reasons beyond management's control. In any case, no one has sufficient information from the 1930's to make a completely accurate estimate of the GNP at that time. The Russian economists did not agree with my guess of $65 billion (a projection based on information gathered by American economists) but they did appreciate the consequences of a much lower figure which would mean a level of poverty for most of the 153 million people in Russia in 1929. After a glass of tea, I think that they recognized the logic and possible validity of my historical figures. They also agreed after some discussion and minor reservations that

made by the Conference Board and a few by the author with the aid of log tables and a slide rule.

The figures for Soviet Russia were obtained from the publications of American economists and reports to the Congress of the United States which were confirmed in discussions with Soviet economists in Moscow.

my figures for 1964 could be about right. Although they don't use the concept of GNP in their calculations, their figures for industrial production, and our joint estimates for services, research, government expenditures, and investments came to about the same round figure of $250 billion. Their only real concern was that this figure was less than half of the $510 billion figure for the U.S.A. They felt that the U.S.S.R. should already be nearer the U.S.A. in total production in view of the fact that its production of steel, electricity, and other commodities is approaching that of the U.S.A.

On an afternoon in Moscow when my guide, Irene, lost her way in the rain and walked us a mile or more to arrive quite wet and half an hour late, it was very rewarding to be met by four Soviet economists armed with records and books carefully page marked and to talk freely all afternoon about the Soviet economy. My interpreter, a relative novice, was helpful, but I found that the economists understood some English and I could follow many of their comments in Russian, and as they said, figures are the same in both languages.

The following day, with a different guide, I met with Professor Shapiro of the Institute of World Economics to talk about plans and projections for the future. I believe that he, too, found my figures for 1964 acceptable, and he immediately identified the source of some of my statistics in the work done by Abram Bergson of Columbia. He did not agree with my projections into the future. In his opinion the U.S.A. will not achieve the annual rate of growth of 3.5 percent that I have used, and he believes that the U.S.S.R. will do better than my projections.

Inasmuch as economists in the U.S.A. seem to think I am too high for Soviet Russia and too low for the U.S.A., I was encouraged to leave my figures as they are.

In the U.S.A. the GNP has increased from $181.8 billion in 1929 to an estimated $510 billion in 1964 in constant 1954 dollars. This increase of 280 percent was at an annual average rate of 3 percent. In current dollars, the actual increase was more than 600 percent.

From 1933, when the GNP fell back to a low of $127 billion in 1954 dollars, to 1964, the total increase amounted to 400 percent, which was an annual average rate of nearly 4 percent. A slightly lower annual rate of 3.5 percent was achieved for the last ten years from 1954 to 1964.

Although it is desirable to establish a trend based on a longer period of time than ten years, and the base year of 1929 was selected because it was a peak year that can be related fairly with the current peaks, it is evident that the trend since 1954 is steeper than the line from 1929 to 1964, and that the line from 1933 to 1964 is even more encouraging. No doubt it would be overly optimistic to project the 1933–1964 rate of growth, but it may be overly conservative to use the longer trend since 1929. For a projection of the GNP into the future it would not be unrealistic to use a figure of 3.5 percent for the annual growth that might reasonably be anticipated.

In any case, the growth in total production and in per capita production in the U.S.A. since 1929 has been greater than ever before in the history of the dramatic economic development of this country. From 1870 to 1900 the economy did grow from $9.11 billion to $37.1 billion in constant dollars (1929 dollars), an increase of 407 percent. The per capita GNP increased from $227 to $488 (2.7 percent annually). The GNP grew at a rate of about 4.8 percent per year but it is hard to compare growth from such a low base with what has happened since 1933. The total growth in thirty years amounted to only $28 billion compared with a growth of nearly $15 billion per year for the ten years from 1954 to 1964.

If per capita GNP is taken as an index, the rate of growth is less startling, but is possibly more realistic and more suitable for comparison with what has been accomplished elsewhere in the world. From 1929 to 1964, the per capita GNP in the U.S.A. increased from $1,492 to $2,680. From 1933, when the per capita GNP fell to $1,007, to 1964, the rate of increase would be nearly 3.2 percent annually. It is important to remember that in 1933 the per capita GNP in the U.S.A. was already far above that in any other part of the world. Even in 1933 it was possible to be out of work in the U.S.A. and to live better than many workers and farmers in Central Europe and the U.S.S.R. As one friend who was working with me in Krimskaya in 1930 put it, after a particularly unpleasant week with little to eat besides eggs, kasha, borsch, and boiled potatoes, "I can live better on handouts in Indiana than on the best that they can give me in Russia."

The GNP per capita has nearly tripled since 1933, and during the same period of time the average income that each person has available on an average to improve his living standards has increased by 262 percent—from $838 to $2,210 in 1954 dollars. This fantastic record of economic growth has made the U.S.A. (by long odds) the richest nation and the wealthiest civilization in world history.

While the GNP increased 400 percent in constant dollars from 1933 to 1964, the increase in current dollars would be more than 1,000 percent—from $56 billion in 1933 to an estimated $600 billion in 1964. (Constant dollars are adjusted to compensate for the changes in prices that occur from year to year and to make them comparable in terms of their purchasing power. Current dollars are the actual dollars received and spent in any specific year.) Per capita GNP in current dollars would be up from only $446 in 1933 to $3,157 in 1964. For the first time, a civilization has attained a level of economic abundance adequate (comfortably) to feed, clothe, house, educate, and amuse all of its members. Enough goods are being produced to justify calling this society affluent. Nevertheless, even now, in 1964, there are still many millions of people who live in drab and crowded tenements and have less than enough food and clothing, although many of them may have old cars and radios and TV sets. It is an incredibly wealthy society, from the point of view of any other part of the world, or from the perspective of history, but it is apparent that its wants and needs still exceed by far what it already has. As John Kenneth Galbraith has pointed out, its collective needs for education, highways, research, etc. are very large and urgent, but these are exceeded by the basic needs of individuals that still remain unsatisfied. To provide for every person and every family only as many material goods as a modest college professor could reasonably require, it is estimated that the gross national product would have to be about three times as large as it will be in 1964.

Even if some method could be found for leveling all incomes, taking from those who have too much to supply those with less than enough, the situation would not be greatly improved. Most of the surplus possessions and acquisitions of the very wealthy would be of little use to those who are relatively poor. Unfortunately, it is not possible to equate surplus income to surplus availability of the goods that are needed.

It is evident that a level of affluence can never be attained for everyone. There will always be some wise or lazy individuals who simply don't want to be affluent if it means working regularly and conscientiously. But an average level of affluence for the lowest income groups is not inconceivable or unattainable, although it would require with the most altruistic management of government and business at least three times the GNP that will be achieved in 1964. Allowing for human selfishness and normal competition and ambition, the GNP would probably have to increase by 400 percent, or more.

Affluence is not a problem in Soviet Russia. Almost no one has much more than a comfortable income. But on the other extreme, almost no one is desperately poor. There are no sections of Moscow or Krasnodar or Krimsk in which forlorn, destitute people live crowded together in dirt and deprivation. If such people exist they are well hidden and there is no reason why they should exist today in Soviet Russia.

Although the average per capita income is only $548, the lowest wage is about $40 per month, or $480 per year, and everyone is employed.

It is difficult to measure Soviet production in dollars because the quality of Soviet goods is not comparable to the quality of similar goods in the U.S.A., Soviet production costs are not calculated according to the accounting standards used in the U.S.A., the value of the ruble is set arbitrarily, and the few accurate and reliable production statistics published overlook many items that are included in the gross national product of the U.S.A.

What can be determined and essentially what matters is that total production of goods and services has increased by as much as 400 percent in the last thirty-five years. Although many consumer goods are still in short supply, the production of goods for household consumption has more than doubled, while capital production has increased by 1,000 per cent. If a figure of $250 billion is taken as the best estimate of Soviet production in 1964 ($1096 per capita), it is estimated that total personal income, or the earnings that consumers have available to live on, is not more than $125 billion ($548 per capita) or about half of the GNP. The remaining $125 billion is still being used to pay for experiments in outer space, other military expenditures, and the development of the U.S.S.R.'s industrial plant.

It is estimated that total production of $250 billion in 1964, which is less than 50 percent of that of the U.S.A., is up from approximately $65 billion in 1929 ($425 per capita). Expenditures for capital goods and military needs are up from $15 billion to $125 billion, and consumer goods up from under $50 billion to about $125 billion. However, it must be emphasized that these estimates are only good guesses. As in the U.S.A. during the period from 1870 to 1930, there has been a continuous shifting of production in Soviet Russia from the home and farm to the factory during the last thirty-five years. The actual increase in production cannot be measured mathematically. It is a difference in kind as well as in volume. The old chimney-oven served well enough for many families as recently as thirty-five years ago. The production of new gas and electric stoves is thus not altogether an additional value, but is in part a substitute value that can now be included in the figures for total production.

In any case, the gross national product has reached a level somewhere near $250 billion, and whatever its quality and kind, it is greater than that of any other country in the world, except the U.S.A. Total industrial production increased by 10 percent per year in the 1950's, with a most recent rate of 8.5 percent. Total annual production increased by nearly 50 percent during the decade from 1933 to 1943, and by less than 40 percent from 1943 to 1953. The increase in production from 1953 to 1963 was about 100 percent. Even after taking into account all of the probable errors and inaccuracies in the available statistics there is ample evidence that Soviet production has increased by 400 percent or more since 1929, despite a crippling war and a continuing social and political revolution.

It is evident that the Soviet economy has been growing at a much faster rate than that of the U.S.A. The economy of the U.S.A. grew at very nearly the same pace during the period from 1870 to 1900, when it was at a very similar stage of development. During that thirty years, the GNP increased by 407 percent, with a per capita increase of 215 percent. This increase compares with a growth in per capita G.N.P. in the U.S.S.R. of 257 percent in the thirty-five years since 1929. In the U.S.A., the growth rate began to slow down at about the time when the average per capita GNP passed the $1,000 level. In Soviet Russia, the economists believe that a similar slowing down will not occur, even though the per capita GNP has now reached a level of $1,096 in 1964. It is very possible that they are right inasmuch as the GNP per capita does not reflect the same living standards that were achieved in the United States at the $1,000 mark. In 1964, the Soviet per capita GNP is about equal to the level of $1,087 reached in the U.S.A. in 1915, and the economy still has plenty of room for further growth.

3

Incomes and Living Standards

S INCE 1929 total personal income in actual current dollars in the U.S.A. has increased from 85.8 billion to about 500 billion in 1964. In constant 1954 dollars the increase would be from $139.2 billion to $420 billion, an increase of more than 300 percent.

In the U.S.S.R. the increase is from an estimated $50 billion to about $125 billion in constant dollars, an increase of only 250 percent. However, if an allowance of at least 25 billion dollars were made for the value of communal benefits, living costs subsidized by the State, the increase would be something more than 300 percent.

On a per capita basis, income in the U.S.A. has increased from $1,140 in 1929, and a low of $838 in 1933, to an estimated $2,210 in 1964. Per capita income in constant dollars has nearly doubled since 1929, and has increased 261 percent since 1933.

In the U.S.S.R., per capita income has increased from $326

in 1929 to $548 in 1964, and no one knows what the low point was during the war in 1942. If an allowance is made for expenses subsidized by the State, the value of the average income of $548 in 1964 should be increased to $782, on the assumption that about 30 percent of the living costs that must be paid out of personal income of $1140 in the U.S.A. are paid for by the State in the U.S.S.R. On this basis, incomes in the U.S.S.R. would be about 240 percent of the 1929 level. However, they would still be substantially below the average per capita income in the U.S.A. in 1929 (about 70 percent).

Incomes have nearly tripled in the U.S.A. since the end of the depression in 1933, and it is true that this society has an affluent look. Moreover, the distribution of income by income levels has changed a little since 1929. Then, the top 20 percent of the population received 54 percent of the income and now they get only 46 percent. The middle 40 percent of the people had 33 percent of the income in 1929, and now they receive 38 percent. (These figures assume that no change in the distribution of income has occurred since 1960.) The lowest 40 percent got 13 percent of the total in 1929, and now they get 16 percent. In 1929, 122 million people shared an income of $139 billion (1954 dollars—only $85.8 billion in 1929 dollars). Only about 5 percent had family incomes in excess of $15,000 and 40 percent had incomes of less than $1,500 per family. The top fifth of the population, about 24 million people, got 54 percent of the income, or $75 billion, more than $3,000 for each person. The middle two-fifths received 33 percent of the total income, about $46 billion, or $940 per capita. The lowest two-fifths had incomes of $368 per capita, or only a little more than $100 per month for an average family of 3.4 persons ($1,250 per year).

In 1929 the economy had touched a new peak of prosperity. Yet, only about two million families lived at what might be called a level of affluence, while about 15 million families, 48.8 million people, lived in poverty with incomes of less than

$1,000 (actual 1929 dollars) per family. It is assumed that a level of affluence amounts to an income of $15,000 or more per family. There was a very wide range in income levels from a low of less than $1,000 per family to more than six million people in families with more than $15,000.

In Soviet Russia in 1929, the spread in incomes was not quite so great, for it is unlikely that there were any people left with incomes of $10,000 or more per family. However, there were still many millions with nearly nothing. The spread was from nothing to what might have been the equivalent of five or six thousand dollars in the U.S.A.

What has happened since 1929? In the U.S.A., Leon Keyserling says that in 1960 the lower two-fifths of the population still had less than "a decent standard of living within the American economy." [1]

He estimated that only seven out of every hundred people lived in families with incomes of $15,000 or more, and 38 million people, "more than one-fifth of a nation" were living in poverty. A comparison with 1929 is very difficult but it is essential in order to get some idea of the income trends that are likely to continue in the future. What is apparent is that in thirty-five years the lower two-fifths have gained only a slightly larger share of the total income, but it is significant that they do have a larger share of a larger average income. The total income of the lowest 40 percent of the population has tripled in the last thirty-five years, as have the incomes of the middle 40 percent. Only the top 20 percent have failed to make proportionate gains. It may be that the lowest two-fifths don't have the kind of living standards that are generally associated with the affluent American society. Only about 10 percent of the population is affluent in fact in 1964. But it is also true that the lower two-fifths are much better off than they were in 1929, not to mention 1933, and are also living as

[1] Leon Keyserling, "Two-Fifths of a Nation," *The Progressive*, June, 1962, p. 12.

well or better than their opposite numbers in Europe or the U.S.S.R.

The progress made in thirty-five years is not as great as it might be, or even as it might profitably be for an expanding economy with more and more productive capacity and more and more goods to sell. But some progress has been made as Table I shows.

Although in 1964 the lower two-fifths (76 million) have only 16 percent of the total income, they do have an income of $67 billion, or an average per capita income of $880 which is considerably better than the average per capita of $368 that they had in 1929. It may be true that about 7 million of these still have less than $350 per capita and another 30 million have only $600 per capita, but if this is so it means that the rest of those in the lowest 40 percent (more than 38 million) have moved up to the $1200 level which was the top of the middle level in 1929. Their share of the total has increased by only a few percentage points, but their incomes, while still inadequate for a "decent standard" of living have increased nearly two and one-half times. Good or bad, this is the trend.

The middle two-fifths, another 76 million people, with 38 percent of the total income, or $160 billion, have an average per capita of $2,100, up from $46 billion in total, or $940 per capita in 1929, an increase of 220 percent.

Meanwhile, the top fifth, with 46 percent of the total, or $193 billion, have $5,000 per capita. This is up from $75 billion in 1929, or just over $3,000 per capita, an increase to a little more than one and one-half times the 1929 level (160 percent).

Although personal income has increased 240 percent per capita since 1929 in the U.S.S.R., not more than 10 percent of the people have even a moderately comfortable standard of living in 1964. It is doubtful that any are truly affluent. Most have family incomes of less than $2,500, or possibly $750 per capita, even after making an allowance for the housing,

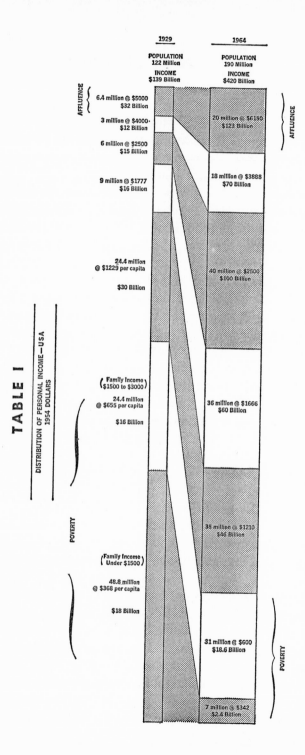

TABLE I

DISTRIBUTION OF PERSONAL INCOME—USA
1954 DOLLARS

1929

POPULATION
122 Million

INCOME
$139 Billion

1964

POPULATION
190 Million

INCOME
$420 Billion

AFFLUENCE

AFFLUENCE

6.4 million @ $5000
$32 Billion

3 million @ $4000
$12 Billion

6 million @ $2500
$15 Billion

9 million @ $1777
$16 Billion

24.4 million
@ $1229 per capita

$30 Billion

{ Family Income }
$1500 to $3000
24.4 million
@ $655 per capita

$16 Billion

POVERTY

{ Family Income }
Under $1500
48.8 million
@ $368 per capita

$18 Billion

20 million @ $6150
$123 Billion

18 million @ $3888
$70 Billion

40 million @ $2500
$100 Billion

36 million @ $1666
$60 Billion

38 million @ $1210
$46 Billion

31 million @ $600
$18.6 Billion

POVERTY

7 million @ $342
$2.4 Billion

TOTAL INCOME—1954 DOLLARS.   N.I.C.B. FROM BUREAU OF CENSUS.

SHARE OF INCOME BY INCOME GROUPS PROJECTED FOR 1964 FROM 1960 FIGURES,
H.J. MILLER, BUREAU OF CENSUS, NEW YORK TIMES   N.Y. TIMES   NOV. 11, 1962.

medical care, and other services provided by the government.

Although there are no published figures available to show the distribution of income by various income levels in the Soviet Union, a rough estimate can be made based on observed living standards. If the upper 5 percent of the population had per capita shares of the goods produced for household consumption of $2,000, a little more than $6,000 per family, and this is probably a high estimate, then these 11 million people would receive a little more than 17 percent of the total goods available, or 22 billion. The 91 million people in the lowest 40 percent of the population would probably receive no more than $30 billion, or 24 percent of the total, and the remaining 55 percent in between would receive the rest. If these guesses are correct, and there is little room for them to be very far off, then the total household consumption in the U.S.S.R. would be divided about as shown in Table II.

TABLE II

ESTIMATED DISTRIBUTION OF HOUSEHOLD CONSUMPTION IN THE U.S.S.R.—1964

POPULATION	PER CAPITA	TOTAL INCOME	SHARE
Upper 5% (11 million)	$2,000	$22.0 billion	17%
Middle 55% (126 million)	576	72.6 billion	58%
Lower 40% (91 million)	334	30.4 billion	24%

The lower two-fifths of the population in the U.S.S.R. are on very slim rations if these figures are correct, but the figures are deceiving. In fact, they must be increased by a uniform amount to cover the rent and other communal services provided at little or no cost, and they must be modified to show that even in the lowest 40 percent of the population, almost no one has less than $1,000 per family. Even the old women who sweep the streets with brooms made of twigs, earn $40 to $50 per month, and since most families include at least two wage earners, it is unlikely that any family incomes would fall below the $1,000 figure.

The lowest income groups in the U.S.S.R. are certainly poor by American standards, but they do share the same kind of living conditions as everyone else. They live in the same apartment buildings, they ride the same buses, and they have enough food to eat. They don't look hungry and their poverty is well concealed.

The average level of personal income has more than doubled since 1929, and nearly all of the increase is divided among the lower income groups. Although there is still a difference of more than 600 percent from the minimum per capita income to the average per capita income of the top 5 percent of the population, there has been a complete shearing off of the highest income peaks and the deepest depressions. More progress has been made toward leveling incomes than would be guessed by anyone who had not examined the available statistics and then had a chance to see the overpowering miles of standard apartment buildings in Moscow with their shops and nurseries, but no garages, and the scientists commuting by Metro or bus.

What has happened is that the peasant who had nothing but black bread and cabbage thirty-five years ago now has some grapes and chickens and a steady job. He and the former kulak who had some pigs and a cow now enjoy about the same kind of living conditions, and both have sons in the university studying to be engineers. And the erstwhile chief engineer is running the local factory and lives in the new apartment building next to the schoolteacher and a young man who works at the gas station. The former plant manager is in Moscow in a staff position for his industry. The changes, modest as they may seem to us in the U.S.A., are significantly in the direction of a more equal distribution of the total income. Incomes have improved, living standards are better, but the belts are still snug in the U.S.S.R. There has been no economic boom, even though the gross national product has increased 400 percent. No one has made a fortune, and very few have attained anything like an affluent standard of living.

In the U.S.A., the trend toward affluence may be exaggerated. The trend is toward a higher standard of living at all income levels, with no significant change in the spread between the highest and the lowest income groups. The U.S.A. still operates on the homely economic philosophy that "money makes the mare go," and motivates economic activity with rewards proportionate to individual efforts and achievements.

The trend may not be toward affluence for everyone, but it seems to be toward a nearly equal opportunity for affluence and the practical results are evident in the changes that have occurred since 1929. In 1929, all families with incomes above $4,500 per year were in the top fifth of the population. Families with incomes of as much as $100 per week were in the top 20 percent, which means that 80 percent of the population had incomes of less than $100 per week (in 1954 dollars).

In 1964, families with incomes of as much as $10,000 per year can't make the top fifth. An income of more than $10,000 or just below the level of affluence, is necessary to make the top 20 percent.

The affluent society has increased from 6.4 million people to 20 million, but it is still only slightly over 10 percent of the total population. Meanwhile, there are still 7 million people in poverty, with less than $1,500 per family, but this number has come down from 48.8 million in 1929. The shifts in numbers within each income group are probably more significant than the percent of total income going to each similar percent of the population.

The number of people in families with incomes in excess of $15,000 increased from 6.4 million in 1929 to 20 million in 1964, an increase of 312 percent. The number of persons in families with incomes between $10,000 and $15,000 increased

600 percent, from 3 million to 18 million. Meanwhile, the number of persons in families with less than $1,500 decreased from 48.8 million to 7 million.

The number of people in families with incomes of $3,000 to $4,500 increased at almost exactly the same rate as the increase in total population since 1929, or 155 percent. All the income groups above the $4,500 level increased faster than the population increased, while those below the $3,000 figure increased more slowly. Table III shows about what has happened at all income levels.

TABLE III

FAMILY INCOME	POPULATION INCREASED: 1929–1964 (in millions)	PERCENT INCREASED
More than $15,000	6.4 to 20	312
$10,000 - $15,000	3.0 to 18	600
$ 6,000 - $10,000	6.0 to 40	666
$ 4,500 - $ 6,000	9.0 to 36	400
$ 3,000 - $ 4,500	24.4 to 38	155
$ 1,500 - $ 3,000	24.4 to 31	127
Under $ 1,500	48.8 to 7	—86
Total Population	122.0 to 190	156

Although the percentage gains in the upper income levels are encouraging, it is still clear that the progress toward affluence is distressingly slow. The rich get richer and the poor get less poor, but there remain 38 million people with family incomes of less than $3,000, and only an equal number with $10,000 or more.

Affluence for everyone is not necessarily the objective of an economy like that of the U.S.A. Nevertheless, the present trends are in the direction of better incomes at all income levels, with an increasing percentage of affluence, and a smaller number of people in conditions of want and deprivation.

In Soviet Russia, the economic rewards offered to and earned by the directors and managers of industry are hardly great enough to motivate the drive and enthusiasm that is encountered everywhere. The trend in living standards is toward a higher level of mediocrity. It is evident that some other satisfaction and reward must move these men and women who are striving to maintain a rate of industrial growth of 8 percent to 10 percent annually. Their success or failure may be a measure of the value of some satisfactions and motivations that can't be counted in dollars or rubles. What is the right name for the satisfaction a man feels in seeing the people in his factory get better houses and schools and a fine nursery where the women who work can leave their children, and a stadium for the young men who play football and win trophies? This is what the director of the *combinat* in Krimsk displayed when he showed me the new apartments and the kindergarten, the parks and playgrounds that have been built in his town since 1944. He directs a factory that produces five million cases of canned food in a year, and that is growing in accordance with the national plan (*gosplan*), but he is not being paid in Cadillacs and caviar for his successful efforts.

The trend in Soviet Russia is toward an increasingly comfortable standard of living for everyone, with a few more square meters of living space for each person, and a continuously larger supply of consumer goods that will be distributed in proportion to the spread in incomes that reflects the varying contributions to the economy that can be made by different people in different jobs.

The basic control on living standards is the nearly equal distribution of living space. In theory, and apparently in fact, every person is entitled to the same number of square meters in the same kind of housing. In Moscow, it is claimed that 90 percent of the population live in apartment buildings that are built to a similar pattern. (The claim of 90 percent is undoubtedly exaggerated. Some of my guides made estimates of only something more than 50 percent. However, it does not make much difference whether the actual figure is 60 percent, 70 percent, or 80 percent. The fact is that most of the people are living in apartment buildings, and that new buildings are being built at a rate that will accommodate all of the population in less than ten years.) The commissar and the day laborer have an equal amount of space and may live in the same building. Eight or nine square meters per capita is the most that can be provided in 1964 and the leveling effect of nearly identical housing for everyone is hard to imagine. A family of three has 27 square meters, about 270 square feet, or two rooms, 10 feet by 13 feet in size, plus a kitchen and a bathroom. It is hoped that this allocation can be raised to 15 square meters per capita in a few years, but as long as the allocation is the same for everyone, regardless of differences in income, and as long as the cost for housing is just 5 percent of the combined family income, the effect is to create what looks like a classless society.

In theory, everyone has an equal chance to move into a new apartment building. The selection is based on need and convenience to jobs, and not on ability to pay. The families living in houses that are torn down to make room for the new apartments have first priority, regardless of their incomes or other circumstances. In practice, it may be true that a man in an important position, or with access to the right people, may be able to advance his name on the list of those waiting for new apartments. However, the significance of such an advantage diminishes as more and more apartments are built, and all to the same specifications. There is little incentive for anyone to want to move if a move means simply that he goes from one apartment that is five or ten years old to another that has just been built. What is more important to the apartment dweller is the convenience of his apartment to his work and to his extracurricular activities. In Moscow, the objective is to create a city without slums and without any comfortable private houses, and the achievement of that objective would seem to be only a few years away.

That the new apartments are not only acceptable but are desirable to most of the Russian people becomes evident after a few conversations with people in Moscow, and after an appraisal is made of the living conditions that have been available in the past.

The apartments not only offer many conveniences that weren't available in most of the old private homes, such as heat, light, gas, hot and cold running water, and plumbing, but they are also convenient in many other ways. There are bus stops and Metro stations located next to or very near the large groups of apartments, there are shops and services, laundries, barbershops, food stores, electricians, and pharmacies on the first floor or in the basement, and in large open areas away from the street surrounded by the apartment buildings there are playgrounds and nurseries and places for outside communal living. Along the main boulevards larger stores, theaters, and movies are spaced so that they are a few bus stops or only a short walk away from almost anyone.

It must be remembered that very little private housing was built in Moscow in the entire period since 1917. Some of the larger homes that were not destroyed were adapted to official purposes, like that used by the Friendship House of the Soviet-American Institute. Others have been made into a number of smaller living units which don't have the privacy or the conveniences of the new apartments. Most of the smaller private homes that have been and are being torn down are little more than cottages and log cabins which may have a few trees and a small garden, but probably don't have running water or inside plumbing. There were no modern homes to be seen anywhere in Moscow like those built elsewhere in the world during the last thirty years and this is of course because there never have been any such homes in Russia.

It is certainly true that a few important people do have *dachas* in the country, and other perquisites that may amount to an affluent standard of living. However, it is an inescapable fact that these advantages go with the job, and are available to the individual only as long as he maintains his position in the political and industrial hierarchy. It is possible that the number of people who have such advantages may increase in the future, but the overwhelming trend is toward uniformity for more than 90 percent of the people.

Differences in family income are not so great as differences in the income paid for different jobs, because of the number of wage earners that will be found in nearly every family. Some families will have refrigerators, TV sets, even cars or motor bikes, and better clothing, but very few could possibly dream of owning a home built to their own specifications.

In the U.S.A. the progress toward affluence in the last thirty-five years is offset in part by the large share of the population that still has less than a decent standard of living. Average incomes have increased from $1,140 to $2,210, but there are still an estimated 20 percent of the population with average incomes of $600 or less. At least 5 percent are at levels of relative poverty and deprivation.

What has happened is that the flying saucer of Soviet income levels has overtaken the lower edge of the broader distribution of incomes in the U.S.A. The great mass of the people in Soviet Russia with average incomes of $548, which might be the equivalent of $782 in our terms, are at a level equal to that of the lowest 40 percent of the population in the U.S.A. Within this group perhaps none are as poor as the lowest 5 percent of the population in the U.S.A., but it is unlikely that very many are much better off than the top segment of the lowest two-fifths of the population in the U.S.A. It is impossible to avoid the conclusion from the statistics that are available and from my personal observation that perhaps 60 percent of the American people have living standards higher than 98 percent of the people in Soviet Russia. At the same time, it is probably true that 3 or 4 percent of the people in the U.S.A. have poorer living standards than almost anyone in the U.S.S.R. The leveling process in Soviet Russia has been more successful than the figures alone reveal. A visit to GUM, the famous department store in Moscow, or a tour of the shops in Kharkov, or Krasnodar, reveals with indelible precision the limited standard of living that is available to anyone, whether or not he has extra rubles. There is more merchandise, but less variety in the shops today than there was in 1930.

4

Population and Employment

IN a modern industrial economy the size and character of the labor force determines to a large degree what economic growth can be achieved. Consequently, the trends in population, available workers, and hours of employment must be known before any estimate of potential growth can be made.

In the U.S.A. and the U.S.S.R. these trends are very similar although they started in 1929 at somewhat different levels.

In the U.S.A. the population has increased by about 55 percent, from 122 million to 190 million; and the labor force has increased proportionately from 49 million to 75 million.

In the Soviet Union population increased by 49 percent from 153 million to 228 million and the work force increased from an estimated 69 million to 102 million.

Since 1929 the total population of the U.S.A. has increased at a continuously faster rate. From 1930 to 1940, population increased by 7.3 percent. In the next ten years, despite another war, from 1940 to 1950 population increased by 15.1

percent, and in the period from 1950 to 1960 the rate of increase went to 19 percent. The percent of the population in the work force has dropped slightly from just over 40 percent in 1929 to 39.6 percent in 1964. In the U.S.S.R. about 45 percent of the population are in the work force and it is estimated that this percentage was even higher in the 1940's and early 1950's during the rebuilding that was necessary after the German invasion of Russia.

Total employment in the United States increased from 47.6 million, or 97 percent, in 1929 to 70.5 million, or 94 percent, in 1964, and it is assumed that all of the work force have been employed continuously in the U.S.S.R. However, the hours worked per worker have come down more in the U.S.S.R. than in the U.S.A. In the U.S.S.R. the work year has come down from an estimated 2,500 hours in 1929 to 2,000 hours in 1964, a reduction of 25 percent, whereas in the U.S.A. the reduction is only 14 percent, from 2,250 hours to 1,940 hours.

After allowing for a level of unemployment of 6 percent in the U.S.A., the total hours worked per worker would be adjusted to 1,824, down 19 percent from 1929, almost exactly the same as in the U.S.S.R. Thus, the total hours worked by all persons employed has increased by 27 percent in the U.S.A. and by only 18.5 percent in the U.S.S.R. (see Table IV).

TABLE IV

MILLIONS OF HOURS WORKED

COUNTRY	1929	1964	INCREASE	% INCREASE
U.S.A.	107,000	136,000	29,000	27%
U.S.S.R.	172,000	204,000	32,000	18.5%

There is a difference in the meaning of extra hours of leisure in the U.S.S.R. as compared to the U.S.A. In addition to enjoying some communal benefits, most workers have time-consuming communal duties. They are members of factory and community committees, they are expected to participate in various ceremonies and social activities connected with their jobs, and a large percentage are in effect required to attend regular classes in industrial and technical schools. The uneducated worker has a duty to complete his education, and the educated worker has the job of teaching and also a share in the responsibility for organizing and improving his industrial or political unit. Very few can escape four or five hours of committed time each week in addition to their nominal hours of employment. The daily chores require more time in Soviet Russia, too. Shopping is slow and difficult, with several stores to be visited almost every day by bus or on foot. It's no wonder that many men and women in Moscow carry a string bag with them wherever they go on the chance that they may come across some of the missing staples or delicacies that they lack.

Although unemployment has been a continuous concern in the U.S.A., where the number of unemployed has varied from a low of less than a million in 1944 to a high of 12.8 million in 1933 and between four and five million in the early 1960's, the available statistics indicate that there has been a similar decrease in absolute employment in the U.S.S.R.

In order to measure unemployment, it is necessary to define what is meant by full employment. Obviously, unemployment has to do not only with the number of people working but also with the amount of work that each of them has to do. Full employment might be defined as employment of everyone who wants to work for as many hours as he would like to work.

Of course, some people don't care about doing a full day's work and others would just as soon not work at all. Possibly, a better definition of full employment would be employment of all persons able to work for as many hours as they should work. Or for the purposes of this discussion, which are to establish trends and make comparisons of the U.S.A. and the U.S.S.R., full employment could be defined as the kind of employment there was in the U.S.A. in 1929 just before the depression or in 1944 when everyone was hard at work to produce all the goods needed for a wartime economy.

If 1929 and 1944 are taken as examples of full employment, then full employment must be defined as nearly 100 percent employment of the available labor force for about 2,250 hours per year. Nearly 100 percent, because even in 1944 670,000 workers were unemployed, or 1.2 percent, and about 2,250 hours, because in 1944 and 1929 industrial workers did work about 45 hours per week for about 50 weeks out of the year, which would be 2,250 hours.

Because it seems inappropriate to include any factor of unemployment in a definition of full employment, the definition to be used here will be 100 percent employment of the available labor force for 2,250 hours per year. (A definition of "labor force" is needed, too, but this is more difficult. The figures used by the Department of Labor will be used here.)

What has happened since 1929 when there was full employment in the U.S.A.?

In 1964 the hours worked per week have come down from 45 to 39.6 and the weeks worked, after allowing for holidays and vacations, have come down to only 49 instead of 50 (this provision for only five additional holidays is conservative) with the result that the total hours worked per year have come down to 1940.4. If this figure is reduced by 6 percent to allow for the number of workers unemployed, it comes down to 1,824, which is 81 percent of 2,250. On this basis there is about 19 percent actual unemployment in 1964.

Unemployment in this sense is certainly an arbitrary calculation. No doubt it could be argued with some justification that a man will work only five hours a day when the ultimate push-button economy is achieved and that all the goods he needs will be produced with ease in that short a time. Ideal employment might be only 25 hours a week for 40 weeks, or 1,000 hours a year. In that case, our present level of employment is nearly double what it should be.

It certainly can't be said that unemployment of 19 percent as calculated here is undesirable or incompatible with the objectives and ideals of the culture of the U.S.A. It is simply a measure of the trend in employment since 1929. During the last thirty-five years, unemployment has varied from 3.2 percent in 1929 to a high of 35 percent in 1935, when the hours worked dropped to 1,830 (50 weeks at 36.6 hours per week), and only 80 percent of the labor force were working, to a new low of 1.2 percent in 1944, and then slowly and steadily up to an estimated 19 percent in 1964.

This trend reflects cultural pressures as well as economic forces and the desire for more leisure as well as automation, but it is a fact that cannot be overlooked or minimized.

What has happened, and will probably continue to happen, is that industry hasn't been able to keep up with its increased ability to produce more goods in less time. To spread the available work it has been necessary to cut down the number of hours worked per man, and to increase the payments per hour. If it had been possible to use increased productivity to increase production instead of to shorten the work week, then the benefits of improved productivity might have been seen in lower prices instead of in higher wages. The degree of success that has been achieved in increasing production is indicated by the actual improvement in living standards that has been accomplished. If industry could have created new jobs faster to absorb all of the workers and all of the work that might have been made available as a result of the increased productivity per worker, there could have been an even greater increase in living standards without any reduction in the number of hours worked each week or each year.

Industry has learned to contain unemployment by keeping

95 percent of the workers employed at shorter hours and with more holidays. It has learned to live with increasing unemployment by spreading the work and by turning up enough new jobs to absorb the yearly additions to the labor force. But it has been scrambling.

It has not yet learned how to use all of the available hours of labor to produce all the extra goods that are needed and could be distributed somewhere in the world. It has not learned how to anticipate what goods will be needed, and to invest in new plants in time to make full use of the available supply of labor.[1]

These trends of population and employment are keys to the future of the American economy. How many people there will be and what they will do will determine whether or not the U.S.A. will achieve the economic goals that are easily within its capacity. What the people can do is important, too. Education, training, motivation, direction, and organization all contribute to the total performance that is accomplished, as do the tools and machines available to be used.

But in the end it is the work that people do that determines the achievements of their economy. In the U.S.A. it could be said that more and more people are doing less and less work. Or it might be said that all of us are able to enjoy a better standard of living without working as hard as we used to. There is certainly nothing wrong with saving work with automation and these figures on actual unemployment aren't necessarily a criticism of the way we are going. They are simply an attempt to measure more accurately what has been happening since 1929.

In the U.S.S.R., none of the labor force is counted as unemployed, although there may be a few gypsies and hooligans who don't work regularly at any steady jobs, and many others who may be temporarily out of work in seasonal industries, or

[1] *Employment and Earnings,* U. S. Department of Labor, Bureau of Labor Statistics, May, 1963.

as a result of breakdowns and other work stoppages. The problem of unemployment does not as yet exist in the U.S.S.R. There is more than enough work to do, and the problem has been to find the workers, and to train them to do the new jobs that are being created in the new Soviet industrial economy.

The trend in employment follows the population curve almost exactly, with some allowance for the losses in manpower during the war, and for the slightly improved standards of living that have reduced the total number of hours worked per capita by a very small but significant figure. More workers are taking time for holidays, and there is some indication that women and younger people are not devoting as many hours to work as they did during the early years of the first five-year plans.

It is estimated that the average work week came down from 47.2 hours in 1955 to 42 hours in 1960,[2] and the forty-hour week is in the offing if it is not already a fact. A forty-one-hour week was the rule in Moscow in 1963, including work on Saturdays ending at 3:00 P.M., but in some industries—coal mining is an example—the week was down to thirty-five hours.

If a forty-five-hour week, with 2,250 hours per year, is full employment, then employment in Soviet Russia must be down to about 88 percent, or 2,000 hours, in 1964. Even this figure is probably high for it assumes that the average work year amounts to fifty weeks, and it is a fact that many workers have holidays of a full month. At least, the Intourist guides do. On this basis unemployment in the U.S.S.R. is about 12 percent compared with 19 percent in the U.S.A. in 1964. Employment is down as much in the U.S.S.R. as in the U.S.A. (see Table V).

In 1929 most workers in the U.S.S.R. worked five days out of six, and often worked more than eight hours a day. A total of 2,500 hours per year is a conservative estimate. A reduc-

[2] Lynn Turgeon, *Comparison of the United States and Soviet Economies* (Hofstra College, 1960), Part I, p. 321.

TABLE V

TOTAL EMPLOYMENT: U.S.A., U.S.S.R.

COUNTRY	1929	1964	
U.S.A.	97%	81%	(83% of 1929)
U.S.S.R.	111%	88%	(80% of 1929)

tion in hours worked to 2,000 in 1964 is hard to prove but is what the Soviet industrialists claim. These figures assume that 100 percent of the work force has been continuously employed. In fact, the level of unemployment due to shifting jobs, shutdowns, and other emergencies might run as high as 2 percent or 3 percent, but probably no higher. There is always some work to be done, and when an industrial plant is temporarily shut down in Soviet Russia, its workers generally find something to do to keep out of mischief.

It is rumored that there is a certain amount of "moonlighting" in Moscow, and that many workers earn extra income by doing jobs in their off hours at premium rates. If the floor needs repairing or the walls need painting, it may be possible to get the work done more quickly and with less fuss by hiring a friend of a friend who may just happen to have some extra paint or linoleum on hand than to wait for a requisition to get through the intricate Soviet channels. It may even be worth the extra cost that is involved, and presumably every Soviet manager has a fund for such items.

The total cost and the total value of this extracurricular production may not be a large percent of the GNP, but it is probably more than enough to offset the percentage of hidden unemployment that must exist.

The trend toward a shorter work week is interesting and significant. No doubt it reflects the increased productivity of industry, some improvement in incomes and living standards, and most of all the inability of the Soviet economy to put new enterprises into motion fast enough to keep all of the available workers busy for more than 40 hours per week. If there were activities under way that lacked workers, the Soviet managers would not hesitate to ask the entire labor force to put in a few more hours. Either the work isn't available, or the trained workers to do it aren't available. This is a trend worth watching, and it will be considered in more detail a little later.

5

Wages, Prices, and Inflation

IN both the U.S.A. and the U.S.S.R. the relationship of
wages to prices reflects the pressure for price inflation in a
growth economy. The worker must devote part of his time to
the building of new factories and the plowing of new fields,
and in both economies he rarely can own the new factories
and fields himself. He is paid for all of his work, what is invested in new capital and what is used to produce goods that
can be sold, and prices rise to offset the imbalance between
the amount he is paid and the cost of the things he can buy.
Or, as in Soviet Russia, his pay is reduced to equal the cost of
those things.

The trends in prices and wages since 1929 show how an
economy must pay for the costs of new capital, for research,
for the exploration of new frontiers, and for destructive and
wasteful wars. The rules are the same everywhere, for capitalism and communism alike. Only the techniques are different, and the resulting distribution of ownership.

In the U.S.A.[1] wages have increased by at least 600 percent since the depression of 1929 to 1933 and prices have increased to absorb the buying power created and to pay for the industrial expansion that has been accomplished.

In some industries for some specific jobs, wage rates have increased as much as 1,000 percent since 1929. Laborers who did heavy manual work in canning factories in 1929 were paid as little as 35¢ per hour. Today they may be paid as much as $3.50 per hour.

But a few specific instances of increases in wage rates don't tell the story. Rates vary from industry to industry and to-day's jobs aren't the same jobs there were in 1929. No perfect measure of the gains made can be found, but if total personal income could be divided by the total hours worked, a figure for income per hour worked would be obtained that could be used for comparing 1929 and 1964 in the U.S.A. and also for comparing earnings in the U.S.A with earnings in the U.S.S.R.

There is no record available of the actual number of hours worked in total by all the people who were working in any single month or year. The figures available are subject to reasonable criticism and suspicion. However, consistent estimates can be made of the average hours worked and the number of people working in any year, and these estimates can be divided into the figures for total personal income that are available to get personal income per hour worked that can be used in measuring trends.

In this way, total income, hours worked, and income per hour worked can be calculated for 1929; for 1933, the low

[1] All of the figures used here are in current dollars and not in constant 1954 dollars because the objective is to look at and measure the inflationary trend. They are all found in government sources: the Bureau of the Census; Bureau of Labor Statistics; Department of Labor; Office of Business Economics; United States Department of Commerce. The emphasis on the record from 1933 to 1964 is dictated by the fact that the inflationary trend started in 1933 and not in 1929.

point in the depression; and for 1964. See Table VI for these results.

TABLE VI

WAGES AND HOURS WORKED—U.S.A.

	NO. OF WORKERS (in millions)	HOURS PER YEAR PER WORKER	TOTAL HOURS (in millions)	INCOME PER HOUR WORKED	TOTAL INCOME—CURRENT DOLLARS (in millions)
1929	47.6	2,250	107	$ .80	$ 85.8
1933	38.8	1,905	74	.64	47.2
1964 (est).	70.5	1,940.4	137	3.64	500.0

Personal income in *current* dollars has increased by more than 1,000 percent since 1933, from $47.2 billion to an estimated $500 billion in 1964. The number of persons employed has increased by 82 percent, from 38.8 million to 70.5 million. Thus, the income per person employed has increased nearly six times in actual dollars which have not been discounted to allow for price inflation. Average income per person employed increased from $1,216 to $7,092. (These calculations are confirmed by the figures for the increase in per capita income from $376 in 1933 to $2,450 in 1964.

The total man hours worked in 1964 are estimated to be 137 billion, with 70.5 million employed for 1,940.4 hours each. This is 185 percent of the total hours worked in 1933, when 38.8 million were employed for 1,905 hours each, or a total of 74 billion hours.

It is possible to use these figures to estimate and compare the earnings per hour of work from 1929 through 1964. However, it must be emphasized that these estimated earnings per hour are not *wage rates* for they are obtained by using the incomes of all kinds of workers and by using an average number of hours worked for all workers. At best they are an average of all the different hourly rates that are earned, and they include interest, rent and all personal income.

They do show that income per hour worked increased from 64¢ in 1933 to $3.64 in 1964.

While hourly income rose by 600 percent from 1933 to 1964, and by 450 percent from 1929, prices increased by 240 percent, from 1933 or 185 percent from 1929. Prices decreased from an index of 100 in 1929 to 76 in 1933, and then increased to 78 in 1934 and to 170 in 1964. In the ten years from 1953 to 1963 the rate of increase, which is the rate of price inflation, was only 15 percent as compared with 33 percent and 55 percent in the two previous decades.

> 1933–43—33% increase in basic prices
> 1943–53—55% increase in basic prices
> 1953–63—15% increase in basic prices

> (Dept. of Labor—1929–56)

In this inflationary merry-go-round, incomes per person employed have increased more than twice as fast as prices and the GNP in current dollars has increased 1,070% since 1933.

Incomes, wages, and prices have kept in good balance, as they must after allowing for the factors of savings, investments, debt, and credit. Price inflation has been followed by wage inflation and both together have accompanied a continued expansion of the economy and an increase in living standards.

Whether or not inflation was necessary to achieve the objectives of the economy since 1933, it is evident that in thirty years inflation has become a habit, not necessarily a bad habit, but nonetheless a way of going that most economists distrust and that managers of industrial enterprises welcome with reluctant satisfaction. If inflation has made profits more certain, and stocks more attractive, and industrial growth more rapid, it has still not become respectable. The practical businessman and the economist agree that inflation is dangerous, and all are fearful of the reckoning that must be faced some time in the future. Yet, inflation has continued for thirty years with assists from government spending, organized labor, and equally organized speculation, and as of 1964, the reckoning is still to come. Despite higher prices and higher wages, and inflated financial values, the economy has survived in good health and continued prosperity. Goods at higher prices are purchased by consumers with higher incomes, and more goods are produced and purchased each year.

The pressures in 1933 for inflation came from workers who needed employment, and from manufacturers whose plants were shut down, and who had goods to sell for less than the cost of production. The problem was to put people to work, and to create buying power for the goods that could be produced. Government pump priming, followed by military assistance to England and France, and then by actual war started the inflationary trend that has continued ever since. The objective of full employment was and still is a major factor in the continuation of this inflationary trend. Continued profits and economic growth are important factors, too, but the most difficult problem is that of keeping everyone employed in an industrial economy dedicated to increasing its productive efficiency through additional mechanization (of its present productive units) and through saving labor.

Improved productivity means using fewer men to operate more efficient machines. The result is to make more men available for new jobs that must be created through continued increases in total production. One answer has been to reduce the number of hours worked, and to increase the hourly wages per man. Such an answer is simply inflation, and in reality begs the question entirely. It spreads the work, and distributes the total production among all the workers, but it fails to take advantage of the increased productivity that could be obtained if all workers continued to work a full day at their present wages.

Another answer would be to increase production, maintain the total number of hours worked, and the current wage rates, and reduce prices in proportion to the increased productivity. More production would thus be available for all workers at prices which they could afford to pay.

Inflation reflects the failure of industry to expand as fast as the supply of available manpower would have permitted. To the extent that manpower saved through improved productive efficiency is not put to work in new productive enterprises, it turns out that it is necessary to increase wages, reduce the hours worked per man, and raise prices.

It is essential to note that with price inflation and inflated wage rates have come inflated earnings and inflated prices for stocks. As capital values have increased, *growth stocks* have sold for as much as forty times earnings. Increased earnings have been a smaller percentage on a larger capital investment, but stock prices have soared, reflecting actual earnings and prospective capital inflation.

Although dividends have been low as a percent of stock prices, the average return on the capital invested in all industry in 1961 was more than 8 percent. The degree to which these earnings were reinvested in industry is a partial measure of the growth that is anticipated and that justifies the high prices for stocks.

What has happened in the last thirty years is that wages have not kept up with higher prices combined with increased productivity and the need for new capital. In total, wages have not increased enough to pay for all of the increased production and the new plants. Profits have increased at as high a rate as earnings (or sometimes at a higher rate), but it would seem that wages and profits together are not high enough to purchase all that is produced. In addition, expanded credit and debt have been required to keep up with and to stimulate the rapid growth in total production.

Public and private debt have increased fast enough to maintain production levels and to keep the brakes on unemployment with the aid of government spending.

At this point what matters is what happened, and not why. Inflation has mounted steadily, the trend toward higher wages continues, and government spending and debt continue to grow. An attempt is being made to relate wage increases to increased productivity, but little progress has been made in finding employment in new enterprises for the total available supply of labor.

Inflation could not occur if goods at higher prices were not purchased with excess buying power made available through higher wages, debt, and profits, but there is no indication in 1964 that any effective measures are available to control the pressures for higher prices, higher wages, and higher profits.

Inflation is not over despite the expressed hopes of most public and private economists.

In theory a monopoly can fix prices at whatever level will return a maximum profit. But the Soviet Union has a double-edged monopoly. It must fix prices to distribute all of the goods that have been produced and to absorb all of the disposable income received by all of the people without regard for profit.

Costs are beside the point in the U.S.S.R. The problem is to distribute the goods and to use up the income. If prices below costs are needed to attract buyers, there's no harm done. Even in the U.S.A. some products and services must be subsidized by the government, and in a monopoly state it is easy enough to offset subsidies with higher prices on other luxuries and necessities.

Prices in the U.S.S.R. can thus be used effectively to regulate consumption, to allocate production, and to tax away from consumers any excess rubles that they may have accumulated.

Wages can be manipulated, too, to distribute goods and wealth, and to create an illusion of growth and prosperity. When all employees work for the same employer, wages and prices can be balanced nicely to get many benefits that go with the distribution of the total available production (no surpluses—no waste). Higher prices can be set for selected commodities to bring in additional foreign exchange, or to hide the actual shortages or unavailability of some other important products. Lower prices can encourage the substitution of one item that is available for another that is in short supply.

Increases in wages, even though they may be offset in part by higher prices, can be used to recognize and encourage workers, and to give them a feeling of participation in the building of an ever stronger economy. Incentive plans, bonuses for extra work, even overtime rates like those in the U.S.A., can be used generously. Free lunches and nurseries, transportation and housing, hospitalization and medical care can all be provided without stint so long as prices are adjusted as required to keep the cost of the available goods in balance with the total disposable income.

What has happened is that prices have increased over the last thirty-five years, and that wages have increased sufficiently faster than prices to increase living standards more than 100 percent. In contrast to the U.S.A. where increased debt and consumer credit have been needed to make up the difference, in the U.S.S.R. total income has kept up with price increases, and some surplus has accumulated to be arbitrarily taxed away on some occasions, or to be stored in new savings banks, or bet on the horse races that now flourish in Moscow!

Even the black market is controlled by the prices set on basic commodities, for the only rubles that can be spent on the black market are those left over after other necessary expenditures have been made.

The history of prices during the last thirty-five years is complicated by several changes that have been made in the value of the ruble, and by the desperate shortages that followed the war in the 1940's. Prices increased from 1933 to 1947, and then fell back somewhat from 1947 to 1959. Prices were reduced from a base of 100 to levels varying from 38 to 43, or by more than half, in April, 1954. Other reductions followed in 1956 and 1957, and on July 1, 1959.

But in December, 1947, currency was devalued at a rate of 1 for 10, wage rates were not changed, and prices were not reduced in proportion to the reduction in the number of rubles available. "Rationing was ended, and new retail prices were fixed at a level intermediate between the previous low 'ration' prices and the higher 'commercial prices.' " [2]

In effect, prices were raised to a realistic level in an attempt to balance purchasing power and the goods at hand to be purchased. Wages were raised, too, and accumulated savings were reduced by as much as 90 percent. (The same wage rates were paid in the new rubles, thus increasing wages by the amount of price reductions which were made on some products in varying amounts.)

In order to double per capita income, wages in total would have to increase twice as fast as prices. This is what has happened in the U.S.S.R. but the path taken is too strange and circuitous to follow in detail.

If prices have doubled, then wages have more than quadrupled, and there has been an estimated increase in real consumption of 230 percent.

In fact money wages in the U.S.S.R. increased 280 percent from 1937 to 1955, from 3,038 to 8,520 rubles. The 460 percent figure *assumed* for thirty-five years is thus not unrealistic. Industrial wages were up 320 percent from 1937 to 1955, from 3,005 to 9,630 rubles.[3]

---

[2] Morris Bornstein, "The Reform and Revaluation of the Ruble," *The American Economics Review* (March, 1961), p. 117, note 5.
[3] Abram Bergson, *The Real National Income of Soviet Russia Since 1928*, (Cambridge: Harvard University Press, 1961), p. 422.

The record of price fluctuations is confused by many factors that cannot be measured with any accuracy. Many new workers have been added to the industrial labor force; many more have been trained to do more difficult and productive jobs; productivity has been increased at the same time by the addition of modern machines and the mechanization of more and more industrial processes. War shortages precipitated other problems, including labor shortages and a two-price system, which was followed by currency reforms undertaken to rationalize prices and to eliminate rationing. Nor can the effect of world markets be overlooked. There is a pressure from outside to put prices and wages in balance with the international trade situation. Prices and wages could be kept in balance by reducing prices instead of increasing wages, but the tendency in the U.S.S.R. has been in the opposite direction just as it has been in most capitalistic countries. The current valuation of the ruble at a price slightly above the U.S.A. dollar reflects this trend. Wages vary greatly even within the same industries in the U.S.S.R.,[4] and their use as a reward for work produced rather than as a device to distribute goods according to need is apparent. They have been used as incentives to stimulate greater production, and it has turned out that an increase in wages is a more effective stimulant than a reduction in prices which benefits everyone equally.

The confusing adjustments in prices and wages and the need for revaluing the currency from time to time reflect a period in which the U.S.S.R. has been scrambling precariously from one problem to another in its attempt to balance the supplies of consumer goods and the available income and to stimulate labor to produce the additional capital goods required. Whether the scrambling is over remains to be seen.

Inflation seems to be inevitable when total wages exceed the cost of the total supply of goods that people can buy.

[4] Robert W. Campbell, *Soviet Economic Power* (Boston: Houghton Mifflin, 1960), p. 135.

Prices rise, and in an economy like that of the U.S.A. a further increase in wage rates generally follows. This is the usual pattern of the inflationary spiral in a free market economy.

What happens in the U.S.S.R.? What happens when total wages are kept at a level nearly equal to the cost of the available goods? Obviously, prices don't rise and wages don't increase unless improved productivity makes more goods available. There is no apparent inflation, and the economy appears to be in balance.

What happens is that wages are reduced relative to the work done and the total production accomplished, in order to balance purchasing power arbitrarily with the available consumer goods. What happens is that the worker is paid for only half of his working time and half of his production. If he received full pay for all of his work, there would be the same type of inflation that is noted in the U.S.A., and considerably more of it.

In the U.S.A. total incomes after taxes are equal to total production of consumer goods plus total private capital investment. Therefore, total income is greater than the value of available consumer goods, and a pressure is created in a market economy for higher prices. Thus prices may rise faster than wage rates even though total wages increase enough to assure an ever increasing standard of living.

It has been suggested that inflation may have served a useful purpose in the U.S.A. by stimulating investment in new enterprises and by providing a hedge against loss. A similar purpose has been served in the U.S.S.R. by what amounts to a different kind of inflation. Instead of offering a potential reward for investing in the economy, the U.S.S.R. has simply demanded that a fixed proportion of all the efforts and abilities of the workers should be so invested. Similar results have been obtained with exactly opposite methods.

In the U.S.A. the worker receives full and fair wages for all of his work. He is then required to pay taxes amounting to as much as 20 percent of his total earnings to cover the cost of government expenditures for public works, services, and defense. Then he is persuaded to invest another 15 to 20 percent in government bonds, corporate stocks, and insurance. Next, prices rise just fast enough to leave him with not quite enough to buy all of the available goods. Finally, he resorts to consumer credit and bank loans to get the funds he needs to buy the remaining items he wants.

In the U.S.S.R. the worker receives wages equal to about 60 percent of the value of the goods he has produced, valued in terms of current prices. Approximately 28 percent of his work is invested by his employer in the building of additional productive capacity, and another 14 percent or 15 percent is set aside immediately for specific projects of the State. Of the 60 percent he receives, taxes take about a sixth, leaving him with less than 50 percent to use in buying the consumer goods available.

The results are much the same. In the U.S.A. the stocks or bonds representing ownership in the expanding economy are not distributed proportionately to all of the workers, whereas in the U.S.S.R. the new industrial economy will belong equally to all of the people. However, in practice the production achieved by the economy is distributed about as equally in the U.S.A. as in the U.S.S.R. All of the consumer goods produced are distributed somehow to all of the people, and the manipulations of wages and prices in both economies seem to produce a sharing of the total goods that is nearly as fair in one economy as in the other.

Inflation, or its reverse, seems to reflect the pressures that exist in an expanding economy. Some way or other *work* for investment in the future must be obtained. Coercion or persuasion, or some of each, may be required, but the need for work that can't be paid for immediately is inescapable.

6

Debt and Government Spending

THE trends in the U.S.A. and the U.S.S.R. over the last thirty-five years suggest that debt is a tool that is needed to finance inflation. In the U.S.A., debt and inflation have progressed side by side, and in the U.S.S.R., the same results have been accomplished in a slightly different fashion.

In the U.S.A., both the words "debt" and "inflation" have a somewhat unpleasant connotation. Perhaps this is because both individuals and private companies have found it possible to get into debt rather easily, and have not always found it possible to get out again. The association of debt with bankruptcies and bad credit ratings is inevitable, and it is not surprising that the evil aspects of debt predominate in the minds of many people. Because private debt makes almost everyone uneasy at one time or another, it is only natural that public debt should be viewed with even greater self-conscious alarm. The feeling that it is wrong for the government to be in debt to the tune of $300 billion is shared by most of the people in the U.S.A.

Nor is there any inclination among businessmen to publicize the fact that private debt is nearly twice as large as public debt. The corporate debt does not make a useful contribution to the corporate image. Burning the mortgage is still a symbol of spiritual purification as well as of financial security. "Neither a borrower nor a lender be" is an ethical precept that gets subconscious recognition even while it is disregarded in practice.

Consequently, the trend toward more and more debt is a clear indication of a force and a need that is being accepted by the economy of the U.S.A. despite the fact that it goes against the grain and the conscience. Debt has increased because it had to. The need for jobs, food, factories, and guns couldn't be put off. Industry had to finance a way out of the depression, the pump had to be primed, and the war had to be won. The money, the capital, and the work had to be borrowed.

Private debt had to be stimulated and public debt had to be condoned because government spending couldn't be avoided. Since 1929, private debt has increased from $161.2 billion to an estimated $780 billion in 1964 in current dollars. Public debt has increased from $29.7 billion to $350 billion.

However, private debt has actually come down in relation to the gross national product from 154 percent in 1929 to about 130 percent of the GNP in 1964. Meanwhile, public debt has increased from 28 percent of the GNP in 1929 to about 58 per cent in 1964. Although private debt reached a high of 228 percent of the GNP in 1933, its trend has apparently been in the direction of equilibrium with the gross national product. (The low was reached in 1945 when it fell to 65 percent).

These trends suggest that a total debt of 180 percent or more of the GNP is what is required to keep this economy in a state of satisfactory growth and prosperity. In 1929 total debt was 183 percent of the GNP and in 1933 it was 300 percent, but it was down to 173 percent in 1950, and to 176 percent in 1960. The total debt has increased slightly from 183 percent of GNP in 1929 to 188 percent in 1964 and has thus kept up with the growing GNP.

It would seem that the factors determining the size of the debt are the rate of growth and the size of the economy. It is significant that public debt as a percentage of the total debt has increased from 15.5 percent in 1929 to 30.8 percent in 1964.

However, public debt reached a high of 66 percent of the total debt in 1945, and it has been on a downward trend for the last twenty years. Apparently, the need for a continuously larger public debt has relaxed somewhat as private debt has done a better job of supplying the funds needed for growth and expansion. Public debt continues to grow in actual dollars but as a share of total debt in 1964 it is less than half as important as it was in 1945.

Few financiers are reconciled to what has been happening for more than thirty years, since the end of the depression in 1933, and most look with foreboding at the prospects for the next thirty-five years. Many profess to believe that the trends are changing and nearly all feel that many changes could be made for the better. At the same time most government economists, and some professors of economics, view the present directions in finance with equanimity, and the buyers of cars, washing machines, houses, and stocks are not unhappy to anticipate paying off their debts with inflated dollars. In any case, the trends are easy to recognize, and no one can deny that they exist.

They are additionally confirmed by what has happened to consumer credit since 1933. From $3.5 billion it has expanded to more than $50 billion. However, $50 billion is equal to only about 60 days of personal expenditures in 1964, whereas $6.4 billion of credit in 1929 was equal to the personal expenditures made in a period of 30 days. Consumer credit has doubled, in fact, since 1929, after allowing for inflated incomes. It has not increased even that much for people in the same income levels. Those with $10,000 incomes today don't owe much more than similar people in 1929. It just happens that a much larger percent of the population is well enough off to afford a few charge accounts and credit cards.

Despite an innate reluctance to go into debt, nearly everyone in the U.S.A. in 1964 is either a borrower or a lender, and many are both. Financial institutions recommend borrowing money on real estate mortgages to invest in securities that will probably increase in value at a faster rate than real estate, and that will provide earnings in the meantime in excess of the cost of interest on mortgages. Such advice has turned out to be right more often than not since 1933 and it may turn out to be right again in the future.

Increased federal spending has been the main reason for the increased federal debt and has provided most of the fuel for the explosive controversy concerning that debt. It was conceded by business that federal spending was necessary during the depression and the war, but spending has continued to rise at a rate faster than the GNP since 1948. In current prices federal expenditures increased 246 percent while the GNP just doubled (from $259.4 to $518.7). It is estimated that expenditures in 1964 will be 90 billion, or 15 percent of the gross national product of 600 billion.

The evil of debt is most clearly perceived in the U.S.A. when it is public debt that is caused by increased government expenditures. What is intrinsically bad is excessive spending by the government, and especially the federal government, and nearly any such spending is automatically felt to be excessive even after it has proved to be necessary and profitable to the economy. Although this feeling persists in 1964, government expenditures continue to mount even though public debt is slowly decreasing as a share of total debt.

A trend as strong and persistent as this must be significant. It has overcome the opposition of industry, Wall Street, the chamber of commerce, and the moral inclinations of most of the people. Yet, the things the government buys are things most of the people want, including defense, social security, space exploration, and better roads.

These are the trends in the U.S.A. Debt continues to grow as the economy expands and a large share of the financial responsibility for the continued growth of the economy is being accepted by the federal government. The fact that debt is not necessarily bad and may even be desirable and good is gradually being recognized. The function and the potential of debt for solving some of the problems of the economy are better understood now than in 1929, and it will be interesting to consider what may be the result in another thirty-five years when the relationship of debt to other factors in the financial mix needed to maintain a desired rate of growth may be entrusted entirely to electronic computers.

In the U.S.S.R., no one worries about debt or inflation. It is not easy for any individual to get in debt, although I was told in Krimsk of a new plan for building cooperative apartments that will be owned by the tenants and will be paid for over a period of ten years. There is also some consumer credit, and savings accounts are permitted and even encouraged. Nor is there any need for debt to finance the activities of business. There is very little private enterprise, and when all the units of Soviet industry operate within the framework of one plan, it is relatively easy to distribute the available funds where they are needed. And there is little need for a public debt when the State owns all of the production in the first place, and can allocate what is produced to serve its own long- and short-range objectives simply by issuing the necessary directives.

In the U.S.S.R. most of the total of all expenditures goes to the government for goods and services supplied by the government. The State has only to retain whatever part of the total it decides to use for new production facilities or for shooting at the moon. It has no need to borrow what it already owns.

The U.S.S.R. has used a pay-as-you-go system to pay for the cost of building a modern industrial economy. With only about half of total production going into goods for household consumption, wages have been maintained at a level about equal to the prices set for that half of the goods. Taxes, especially the turnover tax included in the prices charged for most basic consumer goods, have helped to maintain this balance between total personal income and the products available to consumers.

Thus the capital investments of the U.S.S.R., the costs of nuclear research, the expenditures for defense and foreign aid, and the construction of public buildings, schools, and roads have been paid for by the people of the U.S.S.R. with the value of one-half of their labor for which they received no wages at all! In a capitalist (so-called) economy with a free market for labor this would not have been possible. Instead, all workers would have been paid in full for their contributions to the total production achieved. They would then have been taxed as heavily as the political situation permitted to pay for as large a share as possible of the expenditures made by the government. Finally, they would be persuaded by offers of interest and appeals to patriotism to lend the government the additional amounts needed to pay for exceptional expenditures for research and defense, and in anticipation of profits and dividends, to lend private business enterprises the funds needed to build new productive units and to expand the economy.

Thus, in all fairness and reasonableness it must be noted that the U.S.S.R. has in effect incurred a debt to the people of Russia equal to the amounts in excess of taxes that have been used for research and defense and for investing in an industrial economy.

In effect the leaders of the U.S.S.R. have promised the people of the U.S.S.R. a return on their investment in their new economy in the form of a better standard of living for their children if not for themselves. They have implied that the investment being made will be paid back in full, and with interest, in the development of a new State that will give every worker a comfortable standard of living, with full security against unemployment, and a paid vacation in the mountains or at the seashore. It is difficult to measure this debt in rubles or dollars or to compare it with the public and private debt in the U.S.A. It is a fact, just the same, and it has been sold to the citizens of Soviet Russia in much the same way that citizens of the U.S.A. have been sold on investing their surplus dollars in government bonds or in the stock market. True, the Soviet citizen had no alternative. His labor was confiscated, not merely borrowed. However, he did know what was going on, and he did accept the idea that a heavy investment in the future would benefit him in the end.

If the Soviet defense and industrial expansion had been financed with stocks and bonds in a conventional Western way, the total public and private debt in the U.S.S.R. would be at least as large a percentage of the gross national product as it is in the U.S.A. About half of the GNP has gone into military hardware, new factories, and other government expenditures. With a GNP of $250 billion in 1964, this means that nearly half, or $125 billion, would be represented by public and private debt or by taxes in a market economy. In this sense the debt that is owed to the people of the U.S.S.R. has grown to an astronomical figure from 1929 to 1964, and it is still growing. In 1964 total debt in the U.S.A. is estimated to be about 188 percent of the GNP of $600 billion, or a total of $1130 billion (current dollars). The same ratio in the U.S.S.R. would produce a figure of $470 billion ($250 billion times 188 percent), which seems low in view of the fact that an annual capital investment of at least

$65 billion has been made in the U.S.S.R. during the last ten years, from 1954 to 1964.

It is more likely that a total investment of more than $1,000 billion has been made in the Soviet economy since 1929. This means that the people of the U.S.S.R., who in theory at least are the State, have invested this amount or more in their own productive capacity, in the costs of a war, and in schools, roads, and other public works. They have put half of their labor into building their economy, and have given up half of their earnings for that objective. It is as though in the U.S.A. every person had put half of his income into government bonds or other investments. The Soviet citizen has no bonds to show for his efforts, but he does have an economy with a productive capacity of $250 billion, whatever that may be worth to him in the future.

Compared with the U.S.A., where government spending has reached a figure of $90 billion, or 15 percent of the GNP in current dollars, government spending in the U.S.S.R., which includes investment spending that would be financed privately in the U.S.A., is at a rate of $125 billion or 50 percent of the GNP. Whether it is financed by debt or taxes as in the U.S.A., or simply by confiscation, hardly matters. What is interesting is the trend that can be observed over the last thirty years. Government spending may have continued to increase in total dollars, but in fact it has probably started to decrease as a percent of the total production. In 1929 it is estimated that the government could spend only 23 percent of the GNP, or $15 billion, leaving $50 billion for a minimum standard of living. This percent has increased from 23 percent to 50 percent in 1964 as total production has increased and as living standards have been kept down in order to permit a maximum investment in a new industrial economy (new productive units). Total government expenditures probably reached a high of 55 percent or more in the early 1950's, but seem to have started down again in the last

few years. The costly task of creating many entirely new industries and of modernizing and enlarging other important
basic industries has been done. Further improvements will
be needed, but the longest and most difficult first step has
been accomplished. There is still much to do, but it is possible that the downward trend in government expenditures
as a percent of GNP may continue. It should be emphasized
that there is no indication that the total number of rubles
(or dollars) spent will be reduced, but only that the gross national production may increase somewhat faster than the
amounts allocated to investment and to other governmental
activities.

Government expenditures for purposes other than investment have increased from 8 percent of GNP in 1928, and 21
percent in 1937, to a high of 55 percent in 1944, and back
down to 24 percent in 1955. These expenditures include communal services, government administration, and defense. The
current level of nearly $60 billion per year includes a large
share for defense, possibly $33 billion, and $22 billion for
communal services, leaving $5 billion for government admin-
istration. It is unlikely that the amounts spent for these
items will ever be reduced until an effective disarmanent program is achieved. In the meantime communal costs are rising
and are expected to continue to rise, and no reduction in the
percent of the GNP required for these purposes can be expected in the near future. It is estimated that total government
expenditures are two-thirds as large as in the U.S.A., although
the GNP is only half as large.

In both the U.S.A. and the U.S.S.R., it has been necessary
to borrow the work that was needed to build new productive
units, and to expand the total industrial economy. In the
U.S.S.R., the work was borrowed directly, and in the U.S.A.,
it was financed with stocks and bonds. It is also apparent
that in both economies the amount that can be borrowed to
invest in new capital is limited only by the amount of addi-

tional work that could be done by the labor force. The problem is to borrow the total hours of labor that are available and that aren't going to be used unless they can be invested in new machines and new factories. In practice, the money that is available for investment may or may not be enough to put all of the unemployed manpower to work. The degree of confidence that potential investors have in the profitability of industrial expansion determines the rate of growth and the level of unemployment in a modern capitalistic industrial society like that of the U.S.A.

In the U.S.S.R., it is possible to bypass the borrowing of money from potential investors and to move directly to the employment of all of the excess labor that is available. The problem in the U.S.S.R. is to have the plans completed, the raw materials ready, and the workers trained to do the jobs that are needed.

In both economies, debt might be considered the cost of employing all of the available labor that can't be paid for with the production of additional consumer goods. What is owed is what the worker might have had for his hours of labor if he had been able to employ those hours to produce something that he wanted or needed. It is clear that his hours of labor can't be repaid, and the only return that he can hope to get is in the higher standard of living that his expanding industrial economy can achieve, whether it is the U.S.A. or the U.S.S.R.

7

Automated Production

IN its earlier stages the industrial revolution increased productivity through the diversification of labor and the training of different men to perform different jobs more skillfully than one man could hope to perform all of the different jobs himself. By sorting out and separating the steps in the manufacturing process and by developing skilled labor for each step, it was possible to multiply production many times and to reduce costs.

Once the production of any product had been broken down into a number of relatively simple operations, the next step was to develop tools and machines to assist the skilled workmen who performed those operations, and to replace those workmen when the new machines had learned how to do the jobs themselves. With practice and experience the time came when it was possible to mechanize any production operation if the cost of the required machine could be justified. Workmen were still needed to feed and operate the machines and to check and control the rate of production and the quality of the separate parts produced by each machine. At this point it could be argued that the original concept of the industrial revolution had been realized. There was nothing more to do except continually to improve and refine the mechanized operations already being performed.

It is probably not too much to say that a new phase of the industrial revolution began with the concept of automation. The idea of combining various steps in the production line, and eventually of integrating the entire series of processes and machines that turned out a finished product was, in effect, a step back toward the skilled craftsman who produced a complete product all by himself. The first step in this direction was to link two machines together so that one man could look after them both. Next a combination machine was developed to do both production steps in one continuous operation. Then a series of such machines were connected to permit an automatic flow of the product from one to the next without any break in the chain and with a minimum of human supervision. The final objective is a single synchronized automatic machine that will do all of the operations with infallible accuracy and turn out a perfect finished product!

The integration and wholeness of automated production is important, for it means that there aren't any leftover pieces and parts. Only whole products are produced. There isn't any way to stop production somewhere along the line without wasting whatever is in the machines. The commitment to produce is complete, and the dependence of automated production on scientific, even automated marketing is evident. Once production starts, the process isn't completed until the finished product has been used by the consumer.

The mechanization of production has made it possible to produce more goods with fewer hours of work (per worker) in both the U.S.A. and the U.S.S.R. Its effect on employment and productivity has been much the same in both economies with the result that the U.S.S.R. is reaching about the same levels in 1964 that were attained by the U.S.A. in 1929.

In fact, workers in the U.S.A. in 1929 worked more hours and produced more per hour worked than workers in the U.S.S.R. in 1964. In 1929, the average work year in the U.S.A. was 2,250 hours, and the GNP per hour worked was $1.70. In 1964, the Soviet work year is 2,000 hours, and the estimated GNP per hour worked is $1.22. However, it must be remembered that in 1933, the work year in the U.S.A. dropped to about 1,830 hours, and the GNP per hour worked to $1.20.

In the U.S.A., man's productivity, or the productivity of an hour of his work, has more than doubled in the last thirty-five years. The mechanization and automation of industry has increased the value of an hour's work from $1.70 in 1929 to $3.75 in 1964. In 1929, it is estimated that 107 billion hours were needed to achieve a GNP of $181.8 billion (1954 dollars), while in 1964, 136 billion hours will produce $510 billion of production.

In 1929, there were still many jobs that were done by hand —women packed cans and bottles in cases by hand, ditches were dug by hand, ships and freight cars were loaded and unloaded by hand, hand-pulled rope elevators were still commonly used in factories and warehouses across the country. Since 1929 a continuous stream of new machines has contributed to man's increased productivity and to his release from the back-breaking work he still had to do only thirty-five years ago. The shift in production from the farm and cottage to the factory and from handwork and manpower to the machine has virtually been completed in 1964. This phase of the industrial revolution is pretty well over, and the necessary adjustments in financial institutions and labor practices and marketing methods have been made to fit this kind of an industrial economy.

But now it seems that another phase in the industrial revolution is already well begun. The trend toward automation has moved ahead rapidly over the last twenty years, and is threatening to move man out of the factory and into the control booth. Machines have been created that can supervise, control, and direct the work to be produced by other machines, and there are already some automated factories in which the raw materials are received, stored, measured, and distributed automatically to the processes and machines that convert them into finished products that flow in the correct quantities and varieties to the trucks that will carry them to their ultimate destination. In such factories it would not be correct to say that no human being ever sees what is being done. Men are required to adjust the machines, to make repairs, to watch the control boards, and to develop the plans and directions for the machines. Men are needed, skillful, thoughtful men, but fewer men are required, and their productivity is multiplied many times.

The trend toward automation has moved from the dreams that were already being dreamed as long ago as 1933, to a step by step analysis of the problems to be overcome, and finally to a solution of those problems. In 1964 nearly any process can be performed automatically. Taped instructions can be fed into machines to tell them what to do and when and how quickly. Electronic controls can check every step in a process to assure that it is performed with a predetermined accuracy.

Yet in 1964 there are still many jobs requiring physical manpower in varying degrees, and there are still many processes that have not yet been mechanized. What is most significant is that nearly every machine and every automated series of machines in use could be improved! Changes and innovations are being made continuously and at an accelerating rate, new and better machines are introduced from year to year in every industry, entirely new concepts of production are invented or discovered that require entirely different machines to replace those presently in use, and that may even involve the construction of new factories in new places.

The increased productivity of the men who operate the most nearly automatic industrial units is offset in part by the many men employed in developing and building the new machines, as well as by the large percentage of all workers who are still employed in operations that are only partly mechanized and may still be using vintage machines with only marginal efficiency.

Evidently industry has made little more than a start toward automation in the last thirty-five years. However, its trend is definitely in that direction and would seem to be irrevocable. The die is cast for automation, whose promises have already been tested and realized sufficiently to assure that there will be no turning back.

One of the results of automation will inevitably be increasing standardization of the products produced. The trend since 1929 toward identical products in nearly identical packages is evident to anyone who visits various parts of the country and takes the time to check the supermarkets and department stores. Many of the same brands are sold everywhere, and when the brand names are different, there is rarely any noticable difference in the products themselves. Most items of food, clothing, housekeeping, and houses are standardized already because they are designed to be produced on the same efficient machines and are made of the same most satisfactory and economical materials. Quality standards are uniformly set to satisfy the demands of consumers at the lowest possible competitive price (cost). The same sizes are those which have turned out in practice to be most convenient and acceptable. Even the store layouts and fixtures and merchandising devices are the same everywhere, following a pattern that has been tested and selected to best accommodate the needs and habits of customers.

This trend can continue as long as there are more industries and more customers to automate, but the time may come in less than thirty-five years when there will be nothing left to standardize. There is less variety almost from day to day as more businesses merge and eliminate some products and brands to concentrate on selling larger quantities of fewer styles.

The soap business is the classic example of this trend. New products are introduced on a regular schedule, but the number of items available at any one time is planned to be the minimum required to sell the largest possible volume at the greatest profit. No doubt it is the intention of the soap companies to give all the people what they want as efficiently as possible. The economic effect of such standardization is to make available for more people more of the same goods, with emphasis on quantity and function rather than on quality and aesthetics. In Soviet Russia this trend is called "radical standardization" and is consciously planned with the objective of "producing more goods by reducing unnecessary variety." [1] The advantages of such standardization are greatest in a growing economy like that of the U.S.S.R. Waste can be reduced as straight line efficiency is gained, and the consumer's attention diverted from critical comparison of one product with another. The fact that there is but one product of a kind puts the emphasis strictly on function and de-emphasizes minor differences of convenience and style. A disadvantage is that there is less pressure for improvements in quality, and less incentive for producers to develop ideas that might contribute to a better product. The strongest incentive in a standardized economy is to cut costs. There is some reflection of this result of standardization in the U.S.A. in 1964. Quantity production of standard products at the lowest possible cost has produced legitimate complaints that some items wear out too

[1] *The New York Times Magazine*, March 20, 1960.

fast, are in need of frequent repair, and are made of second-quality materials. In the U.S.A. the consumer can back up his complaints by not buying, and some manufacturer can bid for his trade by producing a better product. For this reason the economic disadvantages of this trend in the U.S. should not be too serious and should correct themselves. The more serious problem is that of sameness, uniformity, and monotony, which can have an effect on the culture of a civilization, and which should be examined carefully in relation to the developing cultures of the U.S.A. and the U.S.S.R.

In the U.S.S.R., the attempt has been made since 1929 to force automation on a relatively primitive industry in accordance with a plan that carefully determined which industries, and which factories in each industry, should be mechanized first. It is evident that the Soviet plan for developing an industrial economy took into account the need for getting as much production as possible from many older and less efficient productive units, and for varying the pace toward complete automation by industries and by units within each industry.

As long ago as 1930 some factories in the U.S.S.R. were as completely mechanized as similar factories in the U.S.A. The trend toward mechanization and automation had begun. A start had been made in converting old factories to new methods, and in building completely mechanized modern facilities to produce entirely new products.

In August, 1930, in Krimskaya, a food canning factory began to operate that was equipped with machines imported from the U.S.A. and Germany. It was mechanized from its ramps with automatic unloading devices for handling truck-loads of raw materials to the end of the canning lines where the sealed cans had to be loaded into crates by hand. The continuous process used surpassed most similar installations in the U.S.A. Yet this was a unique case, almost an experiment, and it operated at about 5 percent efficiency for a day or two before it had to shut down entirely for lack of raw materials. The following year it did somewhat better, and in 1963, the same plant, which had been entirely rebuilt after the German invasion of 1942–43, produced a reported 120 million cans of peas, corn, tomatoes, tomato juice, meats, pork and beans, and other fruits and vegetables. It even produced a small quantity of frozen strawberries.

In 1930, in Krimskaya, a hand-operated sawmill cut planks from logs with a long saw pulled up and down by two men, one standing on a trestle above the log, and the other underneath the log; grain was threshed with a large notched cylinder of concrete pulled in a circle by an old horse; and in a town with a population of 15,000 there were no paved streets. The new cannery was a single unit of twentieth-century technology pushed into an environment of Catherine the Great.

In 1964 a very similar situation still exists in many parts of Russia. There are completely automated units operating as unique experiments or showplaces in industries that are no farther along the road to mechanization than were their counterparts in the U.S.A. thirty-five years ago. It is difficult to determine what progress Soviet industry has made toward complete mechanization, but it is doubtful that it has reached a level as high as that of the U.S.A. in 1933 at the end of the depression.

The GNP of $250 billion in 1964 divided by 204 billion hours worked gives production per hour worked of $1.22, a figure for Soviet productivity that is well below the estimate of $1.70 made for the U.S.A. in 1929.

What is significant is that although no absolute comparison can possibly be made, the U.S.S.R. is certainly at about the same level of industrial development as that attained by the U.S.A. thirty-five years ago. Mechanization varies greatly between industries, and even within an industry. Old plants are kept going regardless of their efficiency because their pro-

'duction is needed at any cost, and the capital isn't available to replace or modernize them. The pressure for rapid industrialization, interrupted and set back by a war, has produced every degree of industrial development in the U.S.S.R., without eliminating as yet many obsolete and antiquated production units. Some very important industries have progressed much more rapidly than the average, and are up to or even ahead of U.S.A. standards in 1964, while many others, including agriculture, still lag far behind.

In 1963 I had an opportunity to see a candy factory in Moscow, a biscuit factory in Kharkov, and a food canning plant in Krimsk (formerly Krimskaya, where I worked in 1930), and I was impressed in each case with the large number of hand operations that are still performed in these modern factories with a very large percent of mechanization. In the same production line with the latest and most efficient German machines for making candy bars, large numbers of smiling women took the finished bars by hand to stack them in old-fashioned machines for wrapping. Most of the operations in the cookie factory, including the mixing machines and the continuous ovens, looked exactly like those in similar factories in the U.S.A., but again the packaging was done partly by hand, and the handling of materials was relatively inefficient. There were no machines for loading cases on pallets, and few automated lift trucks. In the canning factory, cans were being stacked in a storage area entirely by hand. It would be my guess that these factories were operating at a level of efficiency somewhat higher than that of the U.S.A. in 1929. I know that canning lines in the U.S.A. operated at 120 cans per minute thirty-five years ago, and there is little to indicate that similar lines are operating very much faster than that in the U.S.S.R. today. The factories that I visited were probably among the most modern food plants in Soviet Russia, and it is doubtful that the food industry in total is anywhere near the efficiency of the same industry in the U.S.A. in 1929. However, the degree of mech-anization and efficiency achieved by the food industry is not an indication of the progress made by industry in total. The emphasis has been and still is on heavy industry and power, and Soviet steel and electrical plants may be as completely automated as those in the U.S.A.

Moreover, the rate of growth attained, and the momentum generated toward automation is much greater than anything that existed in the U.S.A. in 1929. The idea of automation was a new thing in 1929. Few people had heard the word, and fewer had any idea what it might mean. Russian indus-trialists have the advantage of thirty-five years of experience. They know what it means, and know that it can be done. Moreover, they have their economy in motion. It is already in high gear, and ready to go ahead at a faster pace than the U.S.A. could contemplate at the end of the depression in 1933.

What the U.S.S.R. has accomplished in the last thirty-five years is a total growth about equal to the progress made in the U.S.A. from 1872 to 1933, a period of more than sixty years. If the average per capita GNP in the U.S.S.R. in 1929 was only $425 ($65 billion divided by 153 million), this can be compared with estimated per capita GNP of $384 in the U.S.A. in 1870. (U.S.A. GNP per capita was increasing at a rate of $15 per year.) In 1933 the U.S.S.R. was at about the level reached in the U.S.A. in 1872–73. The per capita GNP in Soviet Russia increased $670 in thirty-five years or somewhat more than the increase of $631 in the U.S.A. from 1870 to 1933 (from $384 to $1,015).

As always, these figures for the two economies are not strictly comparable. However, even with these difficulties it is reasonable to believe that the indicated rate of growth achieved in Soviet Russia does demonstrate a significantly greater momentum than existed in the U.S.A. in 1929 or 1933.

Since 1929, the progress of industry in the U.S.A. and the U.S.S.R. has been interrupted by depressions, crop failures, and war, but in that time the GNP per hour worked has more than doubled in the United States, from $1.70 to $3.75, and has more than tripled in the U.S.S.R., from $0.38 to $1.22.

Most of the credit for this increase in productivity must go to the introduction of better machines and increasing automation. The U.S.S.R. started from a level of industrial efficiency more than fifty years behind that of the U.S.A., and has now regained at least fifteen of those years to attain a level of industrial efficiency very nearly like that of the U.S.A. in 1929. At the present time, the Soviet momentum toward automation is even greater than that of the U.S.A., and it will be very interesting to see what both economies accomplish in the next thirty-five years.

8

Marketing and Distribution

A COMPARISON of the economies of the U.S.A. and the U.S.S.R. helps to make clear what the function of marketing is and how it changes at different stages of economic growth. As per capita income increases from the $326 level estimated for the U.S.S.R. in 1929 to the current level of $2,210 in the U.S.A., the attention of marketing turns from rationing and allocation to advertising and communication, and then to education and some research to find out what new things the consumer could possibly need. The problem changes from one of keeping people alive with a minimum quantity of basic goods to one of helping them to enjoy a more interesting and productive life by making continuously available a broad array of many kinds of goods for every possible purpose.

The U.S.A. has moved since 1929 from an emphasis on selling to an emphasis on consumer research and maintaining maximum retail availability, while the U.S.S.R. has progressed from a situation of desperate shortage to the verge of selling and advertising for a few products. Although it cannot be argued that the U.S.S.R. has reached the stage in marketing that was achieved by the U.S.A. in 1929, there is some overlapping of Soviet experience with that of the U.S.A. Research to measure the needs of consumers is already being discussed in the U.S.S.R., and pricing is being used effectively to control the distribution of many products. Computers are already being used effectively to allocate the supplies of some basic commodities like coal and steel.

However, because Soviet Russia still lags behind the 1929 level of the U.S.A. in the production of most consumer goods, Soviet marketing specialists are still concerned primarily with rationing and distributing the available supplies, while marketing men in the U.S.A. are concerned with making their wares more suitable and attractive to consumers. These differences are reflected in the standard quality of most Soviet goods, compared with the many styles and qualities offered in the U.S.A.

An increase in total production of more than 280 percent since 1929 has created new marketing problems and has made necessary the development of new marketing concepts and tools. In 1929 the idea of total marketing was not yet generally understood or accepted. The marketing department was the sales department, and the function of the sales department was to sell whatever goods might be produced. Most selling was personal selling, and the sales organization was trained to use the skills of argument and persuasion, together with the tools of friendship, entertainment, and occasional blackmail to dispose of the goods it had to sell. The thought that a sale isn't completed until the goods sold are completely used up by the final consumer was heard on occasion, but its significance was not fully appreciated by anyone. The cost of selling was measured very simply in terms of the number of salesmen and the amount of money they had to spend to develop the pressure and persuasion needed to sell the goods that had been produced.

Most sales executives—there weren't any marketing executives in those days—thought of advertising and promotion as extensions of the sales function whose job was to persuade more people to buy. The idea of the "hidden persuaders" stems from this somewhat old-fashioned conception of the entire marketing activity.

Only a few sales executives recognized or dimly sensed that the marketing function was in fact to discover where the goods produced were needed, and to make sure that those goods were properly allocated and distributed to the people who needed them and could buy them. As recently as twenty-five or thirty years ago markets were discovered and measured almost entirely through trial and error methods of selling augmented by advertising and sales promotion.

Prices were determined in the market place by trading and negotiation. If goods couldn't be sold at a price above the cost of production, then the price had to be reduced, and production cut back until supply and demand were safely in balance again, and business could continue at a profit.

It is apparent that such a marketing operation could not cope with the large increase in total production that has been achieved in the last thirty-five years. The rate of growth from year to year could not have been maintained without better methods for allocating and distributing what was produced, and for finding out what should be produced. More efficiency was needed in learning from potential customers what they wanted, and more efficiency was needed in allocating and distributing the goods that were produced to the places where they were required. It is also clear that some method was needed for discovering what consumers would want if new items could be made available. It is now agreed in the U.S.A. that it is marketing's task to answer questions like these. It is generally understood that the marketing organization has a broader and different function than that of the sales department it has replaced, although this concept is not practiced as well as it is preached.

Marketing is not just responsible for selling what is produced, but is responsible for opening efficient channels of distribution, and for finding out in the short term, and also in the long term, what goods are wanted and can be distributed profitably through those channels. It is marketing's assignment to allocate and distribute the goods that are produced (at the best possible price) and to inform the producers what additional goods are needed and could be produced and distributed at a profit.

Thus the objectives of marketing have come to be:

1. Efficient distribution.
2. Efficient evaluation of immediate and potential needs.

To measure the trends in marketing it is necessary to look at the various parts of the distribution system to see what changes have been made, and to note the progress of advertising, market research, and selling in determining what goods are needed now and in the future. It is marketing's job to deliver the goods, and to assist production in making sure that the largest possible quantities of the right goods are available to deliver.

The physical handling of goods from the factory to the consumer, the most costly part of the distribution process with many products, has changed dramatically during the last thirty-five years. From the horse and wagon to the boxcar to the motor truck is only part of the story. "Unitized" loads, "piggy-back" shipments by truck and rail, automatic conveyers, pallets and lift trucks, vacuum systems, specialized over-the-road trailers, and machines for nearly every job of handling, loading and unloading, sorting and packing, stacking cases on pallets, and collecting assorted items in units for delivery to retailers have all contributed to straightening and speeding and cutting the cost of the physical job of delivering the goods. Changes in the construction and operation of warehouses, delivery trucks, shipping platforms, and railway cars have all been engineered to produce additional savings. Experiments in completely automatic warehousing with unmanned trucks moving about on unseen electronic tracks directed and watched on radar from a single control center indicate clearly that the trend is continuing at a rapid pace.

In the area of retail distribution there have been many changes, too. The department store has moved from downtown to the suburban shopping center, and the corner grocery store has been displaced by the supermarket. First, the chain service store shoved aside the old-fashioned, friendly, but costly, grocery store, whose expensive credit and delivery service could not compete with the less wasteful cash and carry operation. Then automobiles and networks of good roads made the supermarket a practical possibility, and the shopping center in the country a natural development. Many of these centers are situated in spacious surroundings with abundant parking space, nurseries, swimming pools, and movie theaters. They include services as varied as TV repair shops, branches of city banks, self-service drug and variety stores, restaurants, hat and dress shops, camera shops, phonograph and record stores, automatic self-service laundries and cleaning establishments, bookstores, and bowling alleys. It is important to remember that these stores and shops could not have been imagined thirty-five years ago. There were hat and dress shops and drugstores, to be sure, but they bore little resemblance to shops of the same name today. Hats and dresses weren't produced and distributed in hundreds of thousands of identical copies in 1929. Mostly they were made partly to order, and often at home. Dressmakers and milliners not only made the products they sold, but to a large extent they made them up.

Today 90 percent of the merchandise in the shops on the plazas is standardized and mass-produced in identical styles and patterns. In the drugstore most drugs were made to order, too, in 1929. Only the old soda fountain faintly resembled the modern drugstore lunch counter. The change in retail distribution extends from the harness shop and feed store, the hardware and the apothecary on Main Street, to the self-stocking shelves and the canned music in the Super, and the moving stairways and tree-shaded walks of the newest shopping centers.

Where will this trend lead the U.S.A. in the years ahead? Coin-operated machines may dispense everything from bassinets to bicycles. Push buttons may eliminate pushcarts in the food stores. A test store has operated already with keys to record the items wanted with a different number for each customer. Retail distribution is many times more efficient in 1963 than it was in 1933, and it continues to make conveniently available a larger variety of all the things that anyone could want.

Yet the cost of retail distribution of most products is still the largest item of cost in the purchase price paid by the consumer, amounting to as much as a fourth or even a third of what he pays. There is still room for improvement, and in another thirty-five years it is certain that many more changes will be made.

The progress made in marketing in the last thirty-five years is often overlooked. The records made in production are exciting, but it should be remembered that these record quantities of goods had to be distributed efficiently, economically, and profitably. Before they could be distributed physically it was necessary to determine just what products should go where, and in what quantities. The problem of allocating production where it could and would be sold at a profit was a new problem in selling that needed the tools of advertising, sales promotion, and research to keep up with the ever increasing supply of new goods. Within limits, costs and prices will define potential markets geographically and by income groups, and it is then up to selling and advertising to find out how much of a product will be purchased by the potential customers who could afford to buy it. Marketing has succeeded in distributing an ever larger volume of goods and in encouraging the production of what could be distributed economically and profitably.

Advertising has found out, quickly and as objectively as possible, what goods could be sold and has helped consumers to select the goods they preferred by communicating the facts about products to potential users of those products. Advertising, as a replacement for, or aid to, selling, has functioned effectively to save billions of dollars and hours of sales effort, and to make possible a rapid adjustment of the economy to the facts of the marketplace.

Advertising, despite its critics, has performed its function nobly since 1929, and deserves recognition and credit for helping to make the growth in production since 1929 a possibility. Advertising was different thirty-five years ago. Then, there was little radio advertising, and no television. Magazines and newspapers were the most popular media, and their share of the total marketing or sales budget was relatively small. The number of advertising messages delivered to the average consumer in 1964 is many times the number to which he was exposed in 1929. Advertising has accepted the responsibility for telling what there is to know about the goods being produced, and it has succeeded in communicating the facts that were needed to assure the continuous growth of the economy. If it has failed at all, it is on the side of too little rather than too much, for there is still unemployment and unsatisfied wants and desires.

The only marketing problem the U.S.S.R. has had to cope with for the last thirty-five years has been that of allocating and distributing the available supplies of all kinds of goods. There have been no surpluses and no need for advertising to let consumers know what could be found in the stores. What was needed was more efficient distribution and better techniques for finding out what goods were most urgently required to keep the economy going and growing.

It was not a question of who wanted or would like what, but of who needed what in order to survive. It was essential to know what consumer goods were needed, first, to keep people alive, and second, to keep people healthy and on the job. It was equally important to find out as accurately as possible what raw materials, supplies, machines, and services were needed to keep the factories running and building, and to prevent wasteful stoppages and delays, and someone had to determine what freight cars and trucks and containers were required to move all of the materials and supplies to the places where they were in critical demand. This was the marketing assignment, to create a distribution system that would function smoothly and efficiently in getting all the products of Soviet industry to their many varied markets and to find out ahead of time what goods would be needed where.

In the 1930's the distribution system in the U.S.S.R. was a continuous puzzle and jumble. People and goods waited for days and weeks to get from one station on the railroad to the next, and the rule was first come first served on materials, supplies, and transportation except for projects given special priority by the government. With the outbreak of war allocations became a desperate matter of first things first, makeshift, and improvisation. Shortages were assumed, and marketing was reduced to finding out what goods were available anywhere, and to inventing ways of moving those goods to the places where they were most urgently needed for the war effort. Even after the war, shortages prevailed. Marketing problems had to do with the statistics of probable production and vital needs, needs for capital goods and materials even more than consumer needs. The marketeers in the U.S.S.R. were the planners who were trying to allocate the total production most effectively for the continued maximum growth of Soviet industry.

As long as there was a wide gap between the available supplies and the basic needs of the economy, arbitrary allocation and planning worked well enough in the Soviet Union. When goods are in short supply, an inequitable distribution of the shortage is hardly noticed. No one has enough. Some have less, but who can tell that? Everything is used and disappears. The planners have succeeded in their jobs of allocating and distributing available supplies.

It is only when the gap closes that errors in marketing calculations become apparent. If too much electric conduit, too many water pumps, or too many shoes are shipped to markets where they aren't used, the surplus amounts become glaring eyesores, especially if the same goods are still in short supply elsewhere. This is the kind of problem that has plagued the Soviet planners in more recent years. The techniques of allocation and distribution have proved difficult to administer without a rational pricing system. Trial and error pricing will move surplus shoes off the shelves, but it won't correct the original maldistribution of shoes. Yet there is no way to find out in the Soviet economy just what the cost of any product actually is or what price would correctly represent its value to potential customers.

This practical problem is evident to the Soviet economists who are responsible for the allocation and distribution of all production, but there is no indication that it is being solved. It seems that the distribution of most consumer goods is still based on income and population statistics and historical patterns of consumption. The use of advertising in increasing amounts indicates that there is some awareness of the need for communication with consumers to let them know that goods they might want are available, but with most items still in short supply there may still be a few years for Soviet market-ing specialists to use in perfecting statistical techniques that can be substituted for rational pricing in distributing consumer goods where they are most needed and wanted.

Although there is very little advertising in the U.S.S.R., the shops and the new apartment buildings in Moscow and elsewhere in the Soviet Union are making increased use of original display techniques. There is imaginative use of neon lettering for store names, and the window displays are bright and attractive. They even include prices that used to be lack-ing, and that suggest that the goods displayed are actually for sale.

Capital goods are allocated without regard to price in ac-cordance with the plans for increasing Soviet production. Marketing is concerned with making sure that all capital goods are distributed on schedule to their planned destina-tions. The problem is one of transportation, goods handling, and warehousing, and has nothing to do with pricing and sell-ing. This is marketing in a seller's market. Each Soviet pro-ducer has simply to tell each of his customers how much he may have of the total production and what the price will be. The price doesn't matter, anyway, inasmuch as the buyer draws a check on the central financial organization for what-ever the amount is, and, in effect, the seller deposits the same check in the same financial institution. It is said that favored buyers get preferential deliveries of larger shares of scarce goods just as they did in the U.S.A. under similar conditions of scarcity, and controlled price levels, and there are expedit-ers and black-market traders who help to short-cut and circum-vent bureaucratic procedures with the unofficial connivance of government agents and central planning authorities.

It remains to be tested whether computers and over-all plans can be substituted for a rational pricing system that would be sensitive to the continually changing needs of a ma-ture industrial economy. In the U.S.S.R. marketing has made progress in straightening and speeding the channels of distribution and although those channels are still crowded and overburdened, an efficient distribution system is well within sight. On the side of measuring future needs and developing potential markets it is not as certain that practical progress is being made. It is possible that the present attempt to make marketing plans on IBM machines will turn out to be a wild goose chase.

An understanding of the function of marketing is as essential to the expansion of an industrial economy as is a complete knowledge of all the techniques of production. The goods that can be produced won't automatically go where they are needed once there is a little more of everything than is absolutely necessary. When per capita income reaches a level of $1,000, the need for marketing skills beyond those of efficient distribution and proper pricing becomes apparent. Some communication is needed to let consumers know what goods there are and what those goods will do. Selling and advertising become increasingly important. Marketing's function is to allocate and distribute what is produced and find out what else is wanted.

It is not marketing's responsibility to determine that what is wanted should be produced. Marketing and advertising, specifically, are often blamed for creating spurious wants for goods of questionable value. Marketing may discover but it can't create the hidden desires for goods that waste the country's resources of materials and manpower nor should it be held responsible for the decisions that are made to produce such goods.

At worst, advertising is an accessory after the fact. Advertising is an extension of selling that is thought to be more efficient but anything that advertising can do, personal selling can do as well or better, though probably at a higher cost. So it is that advertising can extend the sale of a product that is already selling in a small way, but it can't make people buy a product that doesn't have an intrinsic appeal to consumers.

Advertising communicates to many people information that has already proved of interest to a few people. Advertising makes known the facts about products and presents those facts artfully and persuasively so that they may have attention and consideration. The criticism of advertising, then, is that the public should be protected from some facts and that knowledge is not desirable of itself. In Soviet Russia this argument might make some sense. There the State does decide what the people may be told and does determine what the people should be permitted to do. It is harder to justify in the U.S.A. Even if there is waste in advertising, and in the production of products of spurious value, even if billions of dollars are spent to beautify faces that can't be made beautiful, it is still inescapable that many women feel more beautiful as a result of their purchases, and that many men are persuaded against their better judgment that something of value has been added.

Marketing and advertising do have a fundamental responsibility for helping the economy of the United States to cope with the problem of supplying the goods needed to extend a satisfactory standard of living to people all over the world.

The potentials for growth in the lagging economies of the world offer the United States an incentive that includes a promise of continuing profits and economic health and even an assurance of survival. The alternative is to let others, especially the U.S.S.R. and the European Common Market, take these opportunities by default, and to relinquish world economic leadership to those whose aims are oriented to the problems of the economically underdeveloped countries.

All of the marketing skills, including advertising, that have been acquired through long practice and experience in the U.S.A. will be needed to develop new potential markets and to stimulate new international trade.

The importance of advertising for international communication is increasing as the techniques for broadcasting and televising on a world-wide scale approach perfection. The usefulness of advertising for the development of markets everywhere in the world is just beginning to be recognized.

Consequently, it is important to understand what advertising is. Advertising is essentially a form of communication. It is not entertainment. Its purpose is not to amuse but to inform. It may be pleasant and interesting, but its objective is communication, not edification or recreation. It may use a mood or an association to create an atmosphere which aids communication, but it must avoid creating an atmosphere which is confusing or distracting.

The idea that advertising must be sugarcoated because it is unpalatable and unwanted is basically unsound. Advertising is wanted and is palatable when it performs its function of supplying information.

The future of advertising is nonetheless in doubt. The trends toward market research, product standardization, and greater marketing efficiency all contribute to a reduction in the need for advertising to inform potential users of the merits and advantages of specific products and brands. When industrywide research measures with statistical accuracy the qualities and performance of all sizes, styles, types, and models of a product and makes all the facts available to potential customers, the need for advertising the brands of such a product is reduced to reminding customers of the brand names available, and to attaching pleasant connotations to those names. The job of advertising will begin to be that of *conditioning* potential users to prefer and accept one brand in preference to another. Such advertising can be useful to an industry only to the extent that it creates total usage of the industry's products greater than would have existed without the advertising. Consequently, as products become standardized and efficient distribution makes them available to all potential customers, and as advertising and sales promotion make all consumers aware of all the brands and styles there are, the possibility of increasing industry sales through additional advertising becomes less and less likely.

In Soviet Russia the effect of rational standardization may change and limit the usefulness of advertising. As long as there are fewer kinds and styles of each product, there will be less to say about the different kinds. Once the standards are described and understood, there will be little more to communicate. It is even possible that the U.S.S.R. will be satisfied with a less wasteful standard of living than that desired in the U.S.A.—less wasteful, less stimulating, and less beautiful—but that is far from certain.

Advertising does have less and less to say as products become more and more standardized. Or it could be that selling has less to do and that market research and well-directed information to the various levels of distribution and to consumers through advertising can combine with automated physical distribution to do the complete marketing job.

But there is still another trend that must be weighed and considered. This is the trend toward longer-range market research to find out what products will be wanted five or more years in the future. As market research becomes concerned with all markets and with total marketing activity and discovers techniques for measuring total needs in relation to total production, it will learn how to predict with accuracy what kinds and quantities of goods can be marketed at a profit this year, next year, and the year after. It will thus be able to guide the economy toward the production of the goods that are needed and wanted and away from excess production of some products and underproduction of others.

It is very possible that market research will team with quality controls and standards to improve the efficiency of advertising communication in the U.S.S.R. and to eliminate some of the waste that cannot be avoided in a competitive economy. Soviet Russia can develop its marketing function based on its knowledge of the progress in marketing that has been made in the U.S.A. in the last thirty-five years. If it is at a stage about level with that of the U.S.A. in 1929 and can learn now what the marketing specialists in the U.S.A. have found out since that time, it can organize its marketing activities to achieve an efficient distribution of its increasing volume of production in accordance with its over-all program for industrial expansion. It is very likely that it will use more research and less advertising.

The concept of marketing includes the idea that markets can be measured, and that it is possible to ascertain in advance of advertising whether a product is acceptable, and how many units can be distributed at a given price. The probable profitability of a product can thus be determined before the risk of producing the product itself is taken. Market research has grown steadily in consequence and respectability during the last decades. It first set out to gather and organize all statistical information available and then it started to dig into all the other factors that could influence a marketing decision. Step by step, ways were found to measure the needs, likes, preferences, and habits of consumers, to measure the patterns of behavior surrounding a buying situation, and to analyze the motivations back of the consumers' choices and selections. Products, packages, labels, advertising, and promotions have all been researched, and market research, too, should have a share of the credit for the total marketing job that has been accomplished.

The trend has been toward more and more research, and the publicity given to some questionable techniques of motivation research has helped to lead astray the economists who assume that it is true that motivations can be created to assure the purchase of a worthless and unnecessary product. Nevertheless, in 1964 all the research together is a relatively small portion of the total marketing cost, and it is still able to solve only a few of the more elementary marketing problems. It is just making a start, but the fact that it is growing and that it is concentrating its attention on getting more and more marketing data is significant for the economic future.

From automated warehousing and transportation to market research and pre-tested advertising it is evident that all of the marketing trends have been moving in the same direction, toward mechanization and automation. It is hardly a guess to predict that these trends will lead to highly sophisticated mathematical marketing operations within a few more decades, not only in the U.S.A. but also in the U.S.S.R.

9

Scientific Management

ALTHOUGH both the U.S.A. and the U.S.S.R. profess to believe in scientific management, they don't necessarily agree on what scientific management is. A business executive in the U.S.A. would argue that "scientific" management is impossible when prices are determined by a central authority without regard for the pressures of supply and demand in a free market, and a Commissar in the U.S.S.R. would feel just as strongly that "scientific" management precludes the siphoning off of excess profits to a privileged minority. Both sides claim to have scientific economic sanction for their different moral and philosophical beliefs.

Obviously, it doesn't matter to pure economics how incomes are distributed or how prices are set. Economic calculation has to do with verifiable data, not primarily with beliefs. That every man should have a nearly equal fair share of the total production is not a scientific statement in this sense. Capitalism and communism aren't solely economic concepts. They involve, to varying degrees, political and moral values.

Nevertheless, both economies have been moving ineluctably toward scientific management. Both find it impossible to avoid using all the data they can find.

Over the last thirty-five years there has been a continuous trend in business away from reliance on experience and rules of thumb toward the use of mathematics and research and slide rules. In the U.S.A. hardly a business manager in 1964 has not taken part in at least one or two study groups or academic seminars somewhere in the Adirondacks or the Rockies. The list of textbooks provided for top-level executives is all inclusive, and covers every possible aspect of production, marketing, and finance. Hundreds of business associations offer training courses, scholarships, internships, speeches, essays, and radio and TV scripts on every conceivable subject. The swing from suspicious and reluctant acceptance of any aid from the brain trusters in 1933 to frantic pursuit of the most esoteric and theoretical dreams of the economists has been nearly complete. In today's planning conferences, operations research and computers are dragged in, whether they are needed or not, and the tendency is to err in the direction of too much research and the acceptance of theoretical answers even when these violate common sense. The exchange of ideas and methods by business executives in classrooms has gone so far that whatever any one of them discovers is known almost immediately to his competitors. In fact, the technical experts of one company often will volunteer to help a competitor to solve a production problem with which he has wrestled unsuccessfully.

The trend in most industries is toward standard techniques and tested methods. Automation of production processes in the plants has extended into the offices of production management. Managers with a genius for improvising shortcuts and improvements are of little use when the automatic machines dictate what can be done. Any change in the programming of production becomes a problem for a production committee and may well require the assistance of outside scientists and technicians in the employ of suppliers, consulting firms, or trade associations.

The trend is clearly toward management that has available the facts needed to make scientific decisions and that prefers decisions based on facts to those relying on opinions and inspiration.

Although this trend is most apparent in the area of production, it is equally at work in the fields of finance and marketing. The skills of the accountants and comptrollers have changed since 1933 in a direction that is as significant as the increase in debts and government expenditures. Not only are more records kept in 1964, but more use is made of the available records than was dreamed of thirty years ago.

The bookkeeper on a high stool with a quill pen may seem antique and right out of Charles Dickens, but he persisted in fact longer than many financiers like to remember. Automated finance got a late start, but it has made enough gains in the last decades to wipe out any handicap it had at the beginning.

The old-fashioned bookkeeper kept an accurate record of what came in and what went out of the till, and of what was due and what was owed. He drew a simple numerical chart of what happened, after it had happened. Without any income taxes or payroll deductions or sales taxes or subsidies and fringe benefits, his chart could be forthright and simple, and when it was done, he was through. His boss was the treasurer or the president of the company, who read the chart and determined what should be done to cut costs or increase income or get new capital.

Today the accountants maintain a running record of the cost of every function of every department in the business. The factors of cost are related one to another in various combinations and their interdependence and influence is noted mathematically and is used to project cost trends and potentials and probabilities.

The comptrollers and financial vice presidents who direct the accountants and understand what they are doing explain it all to the presidents, treasurers, directors, and management committees. Their contribution to scientific management has increased consistently over the years because they have been able to accumulate knowledge never before available and have demonstrated its usefulness.

Progress toward science in marketing is clear, too, in measurements of product acceptability, advertising and sales efficiency, and market potentials for every kind of product. Opinion has less and less to do. Progress toward science is progress away from guesswork and trial and error.

So the trend toward scientific management is a trend toward less guesswork in managing the economic affairs of the U.S.A. It is an inescapable trend when the availability of more and more information makes guesswork unnecessary in more and more management situations. Yet, it is a trend which many managers regret for it reduces the premium that can be paid for luck and inspiration and political acumen. The manager with guts and drive fears that he may be replaced by a cold scientist or even by a mechanical robot capable of producing mathematically correct answers from a complete analysis and evaluation of all of the relevant data.

The trend toward science is a trend toward knowledge and away from intuition, toward facts instead of opinions, toward logic and mathematics to replace daring and enthusiasm. It is a trend that seems to be gaining momentum throughout the economy and if it continues it will account for significant changes in the next thirty-five years.

This does not mean that individual business managers in the U.S.A. are being overpowered by computers and operations research. This is far from the case. Experienced executives are not overwhelmed or overawed by the possibilities of reaching better decisions with the aid of more data and more mathematical computations than they could afford to make in the past. As yet, they haven't found that the new scientific techniques have relieved them of any of their responsibilities or of any of the time required for adding reason and judgment to the cold facts.

During the last thirty-five years the economy of the U.S.A. has been moving from independent personal business management based on experience, negotiation, and opinion to impersonal and objective scientific management based on facts. It has been moving from free enterprise with many separate small units competing freely in an environment without any serious legal or economic inhibitions toward a controlled private enterprise made up of fewer and much larger units that compete within a framework of increasingly complex and restrictive laws, and self-imposed economic limitations.

In the U.S.S.R. the trends have been in a different direction. In the 1930's the State ran all of industry and agriculture directly from Moscow. Each industry had its central planning and administrative organization and the local managers of most individual factories could make no policy or important operating decisions without approval from headquarters.

Central planning and decision making was intended to be a step toward scientific management. Decisions were not to be based on local pressures or the subjective experience and opinion of managers who could have access to only part of the facts. All decisions were to be weighed and examined in the light of all the available facts, and were intended to be objective and in the interest of the total economy.

In practice the complete transition from practical local decisions based on experience to centralized scientific theory proved to be inflexible and inefficient. Local managers were afraid to do anything. To keep their jobs and remain physically healthy they found it was wiser to wait for orders.

In Krimskaya they never did give the signal to start the new corn canning operation. They simply waited until the Americans lost patience and took the matter into their own hands, turning the switches and calling the workers to their jobs. As a result, a move back toward regional decentralization was undertaken although the final responsibility still remained with the central planning and administrative authorities for all industry in Moscow.

The idea of an economy that is coordinated centrally has persisted, although considerable progress has been made in delegating responsibility and authority. Says Robert Campbell in *Soviet Economic Power*: "The director of an enterprise is held accountable for performance of the enterprise, but at the same time he has authority to direct the affairs of the enterprise commensurate with this responsibility. . . ." [1] There has been a swing back from what was considered scientific theory to a more workable program for getting things done within the over-all plans.

In Krimsk, I was amused when the director of the factory made it clear to me that he had no boss in Moscow after I had mentioned the name of a chief specialist who had talked to me in Moscow about the canning industry. The director asked me who my chief was in the U.S.A., and after I had told him, he emphasized again that he had no such chief. In his *combinat* he is the director, but he tried to explain that there are no chiefs in the Soviet organization, that everyone works together to accomplish the over-all plan (*gosplan*). In any case, I had a difficult time in getting to see him in the first place, and it was apparent that he had the authority to say whom he would see, regardless of any recommendations he might get from Moscow. That he did see me at all was probably due to the persuasiveness of my Intourist guide in Krasnodar.

Both the U.S.S.R. and the U.S.A. have more or less scientific

[1] *Op. cit.*, p. 118.

management in 1964, but this does not mean that they have the same kind of management or that they agree on what the science of economic management is. The science of economics is still not an exact science; many economic theories have still to be tested, and about all that the U.S.A. and the U.S.S.R. agree on is the possibility that there is such a science. This may seem to be a small area of agreement, but in fact it is quite meaningful and significant. Because if economics, like physics, is a science, in the end its postulates and theorems will be proved as far as possible and both the U.S.A. and the U.S.S.R. will have to accept the same economic hypotheses and principles. In the field of nuclear science the U.S.A. and the U.S.S.R. agree on the rules of the game. Neither state questions the recent discoveries in the field of physics, though each tries to find more effective ways to use this information, and there are still differences of opinion on questions of time and space that have not yet been solved.

The science of economics has been hindered by the conflicting theories attached to different political ideologies. Interested advocates of these theories have limited and delayed scientific analysis and research. Economists have taken sides and have set out to defend their positions. Consequently, the science of economics is still in a stage of exploration and definition. The U.S.S.R. continues to believe in some Marxian hypotheses, but it is already testing some capitalist techniques. The U.S.A. believes in capitalism but is continually experimenting with programs that seem to be in the direction of socialism. What is relevant here is that both are testing new ideas in practice and that both are moving toward answers that will be based on pragmatic results and not on treasured beliefs.

Economics will be more scientific when economic decisions are based more on tested information and less on personal or political opinions. If both the U.S.S.R. and the U.S.A. continue to add facts and subtract opinions, they should reach the same conclusions in the end. A trend in this direction has been continuing for several years as both states have made progress in finding and organizing information.

The U.S.S.R. has never denied the need for information, despite its confident belief that the facts, when obtained, would reinforce the economic theories of Marx and communism. The scientists of the U.S.S.R. have proved in other fields that they know how to identify and use knowledge that is at hand, and it is unlikely that they will shrink from the meaning of facts in economics when these are there to know.

One of the most significant trends in the U.S.S.R. in the last thirty-five years has been the progress made in finding out and recording statistical information about the Soviet economy. Very little useful information was available in the 1930's (less was available in the U.S.A. than most Americans remember!) and a long and strenuous effort to collect as much data as possible was interrupted during the war years and has only recently produced series of comparative figures that have significant meaning. The volume of information that is becoming available is increasing at a rapid rate and is now being fed into electronic computers to be digested, analyzed, and used for future decisions. Thus, the U.S.S.R. is moving toward scientific management from a different starting position than the U.S.A. but in a remarkably similar direction.

10

Meaning of Trends in Typical Communities

ALTHOUGH Krimsk in 1964 is still a small country town, it has changed so completely that it is almost impossible to find in the town or the surrounding countryside anything that looks as it did in 1929. Then, Krimskaya was a village of 15,000 people, with no paved streets, and consisted mainly of small cottages with thatched roofs. Near the railroad tracks the L-shaped canning factory dominated the landscape as well as the town's industry. Unkempt, barefooted children ran about the filthy streets. In the summertime, when it was not necessary to have a fire in the houses, cooking was done outdoors in ovens made of plaster. Sunflowers, the main crop in the village, could be seen drying on the thatched roofs. Near the center of town was a small square surrounded by the fenced-in houses of the kulaks who had their own small orchards and gardens. Transportation consisted mainly of horse-drawn vehicles, though most of the workers walked to the factory located a mile away from the village square.

In 1964 only the canning factory looks vaguely the same, though it was entirely rebuilt after the German invasion in 1942. Two tall chimneys now rise above it, and a number of other additions have been made. Today the population of Krimsk numbers 30,000. Many of the old houses now have running water, as well as gas for cooking and heating; and there is inside plumbing in all the new buildings. The town is clean today; the children wear shoes and are nicely dressed. The area between the old village square and the factory is now completely filled with new apartment buildings set among paved streets with sidewalks and pleasant shade trees. There is a new nursery and a kindergarten, and a three-story school building. Flower beds appear everywhere—along the new parkway and in front of the *combinat* office. There is now a theater where the square used to be, and nearby a second small canning factory that processes food products for local consumption.

The town has two large parks, each with its own stadium; and there are playing fields for the increasing number of young people who are interested in sports, and for competitions between the teams sponsored by the *combinat* and those of other factories in the area. The director of the *combinat* told me with pride that his teams had won first prizes in most of their games for a number of years.

The horse-drawn carts of 1929 have for the most part been replaced by a modern bus system, but at present there are few privately owned cars. The Krimskaya that I recall has disappeared. Nevertheless, I am sure that the people that I saw in the present-day Krimsk were the same I had met thirty-five years ago. They have new living standards and new attitudes toward living, but their character of combined stubbornness and good humor seems to be the same.

The road from Krimsk to Krasnodar is another change that is hard to believe. In 1929, the journey from Krimskaya to Anapa on the Black Sea could be made by car, but it was a hazardous trip on dusty country roads that were little more than cattle trails. It was not considered safe to take the back roads from Krimskaya to Krasnodar. Now there is a two-lane highway that follows the railroad and that is not unlike the roads in Michigan in 1930. It is already being widened, new bridges are being built, and no one can remember that the only safe way to get from Krasnodar to Kimskaya in 1929 was by railway, a trip that took several hours in crowded third-class carriages. Today the trip takes only an hour and a half by car. In those days, the railroad tracks ran along through acres and acres of sunflowers, and there was no road at all to be seen. Now the highway passes through fields of corn and castor beans, and by newly planted orchards, small vineyards, and berry patches, and crosses irrigation canals with neat rows of poplars planted along each side. There are a number of small, pleasant villages, with unusual fences made of clay and plaster, and there are cement factories and oil wells to be seen from time to time on the side toward the railroad. The land is rich and well cared for; there is even modern farm machinery to be seen, and instead of the single crop of sunflowers, the farms now produce a variety of crops, although sunflowers are not forgotten.

Krasnodar was founded in 1794, about a hundred years before Kalamazoo was settled by adventurous pioneers from Holland.

Kalamazoo has changed, too, since 1929, but it is still recognizable as the same place that it was then. Many buildings are still the same on the main street, including the old Burdick Hotel, and many blocks of old houses are virtually unchanged. The broad square has changed very little, although some streets have been blocked off to make a shopping plaza, and there are a few new buildings. Most of all, the friendly, conservative feeling of Kalamazoo is just the same, although there has been added a note of confidence concerning the future that was not there in 1929.

Kalamazoo now has its own airport, but it has gained even more from the two new superhighways that run east and west, and north and south through the town, and that provide easy access to all of the major markets in the Middle West. Chicago is now only three hours away by car, whereas I recall that it used to take most of a day to drive from the north side of Chicago to Kalamazoo in the twenties.

Although celery was the chief crop in the area near Kalamazoo in 1929, only a few acres of celery are cultivated there today, and most of the old marshes have been converted to the growing of pansies for home gardens. Many thousands of flats of pansies are now marketed in the shipping radius of Kalamazoo. In addition to the cultivation of pansies, in 1964 the city is better known for its pills, paper, and peppermint. It is the headquarters for one of the largest pharmaceutical companies, the Upjohn Company, and the paper industry is well represented with four or five important companies, making paper and all kinds of paper products. Another local concern, the A. M. Todd Company, specialized long ago in the production of peppermint extract, produced from mint that was grown locally. Kalamazoo is still the center for the production of peppermint flavoring, both real and synthetic, but the mint that is used for the genuine product now comes mostly from Oregon. Because of its fortunate situation, and the new roads that lead quickly to the major markets of Cleveland, Detroit, Indianapolis, Chicago, and Milwaukee, Kalamazoo has prospered as a center of industrial production. The city itself has grown in size from 50,000 to more than 80,000, and the county has grown from about 90,000 in 1929 to 183,000 in 1964. As in Krimsk the population has doubled, and has increased at a rate faster than the increase in population in the United States as a whole due to the trend toward urban industrialized areas and away from the farms and villages.

Kalamazoo College, founded in 1833 and originally a Baptist institution, is widely recognized for its high scholastic standards in the sciences. It accommodates 900 students and includes two semesters of study in Europe as part of its regular college program. Consequently, it has a waiting list of alert young people who wish to enroll. Western Michigan College, with an enrollment of 12,000, is primarily a teacher's college. It is now embarked on an ambitious expansion program that is needed to keep up with the increasing number of students expected in the years ahead. Many new multistory buildings are nearing completion to provide additional classrooms and study halls, and there are also new dormitories and attractive apartments for married students that rent for only $35 per month. All the buildings are modern and functional, not too unlike the best modern architecture in the U.S.S.R., and they will soon fill the new hillside campus that is just now being landscaped and planted with shrubs and trees. Other new high schools and grade schools may be seen in nearly every section of town.

In Krimsk and Krasnodar the schools are simpler and more austere, but they are used around the clock. Even in 1929, before the factory was completed, a one-story temporary building was used as a school for the factory workers, and now there is a long three-story building, only a little more pretentious, in almost the same location. It is busy most of the time with as many classes for adults as for young people of high school age.

One striking difference between Kalamazoo and towns in the U.S.S.R. is in the number of churches. Perhaps there is a church in Krimsk, but I didn't see it. There was one in 1930, but it was closed and its priest sent away during my visit that summer. The large Greek Orthodox church in Krasnodar was closed, too, at that time, but in 1963 it was open, regilded and refurbished, with regular services for a small number of worshipers. In Kalamazoo, there are churches everywhere, from the handsome Episcopal church near the square and four Catholic churches to the many small Dutch Reformed churches that sprinkle the town. Not so long ago the old Dutch families went to church most of every Sunday and often during the week. Their puritanical fervor is reflected in the stodginess and conservatism that shows through in the older sections of Kalamazoo.

There are seven golf courses in Kalamazoo, three private country clubs and four public courses. As elsewhere in the U.S.A. sports are becoming more and more popular; and fishing, hunting, boating, and skiing are an important part of everyday living in Kalamazoo. With lakes and hills and woods only a few minutes away, almost everyone can participate in some kind of outdoor recreation. There is a similar interest in sports in Krimsk, but the emphasis is on team sports. Golf is still considered a capitalist game and few individuals can afford to arrange private hunting and fishing expeditions. They do make weekend trips to nearby towns to see their teams in action, and it is particularly easy for people in Krimsk to get away for a swim in the Black Sea. Although they don't bring their horses along to Anapa as they did in 1930, and do wear swimming suits (many didn't in 1930), they all believe in enjoying the sun and the sea as frequently as possible. A continuing interest in drama, ballet, and music is evident in Krimsk and Krasnodar. Even in 1930 our cook's daughter was loudly applauded in a performance of a Chekhov play in Krimsk; and in 1964 there is a new theater in Krasnodar that presents attractions from Moscow and even from other parts of Europe. It draws an audience from all of the Krasnodar area. In Kalamazoo the Civic Auditorium and the Art Center are quite new, but the interest in ballet is probably not quite so intense. There is a local symphony orchestra with a paid director, and the college drama clubs offer an interesting variety of modern and classical fare that is presented enthusiastically, if not professionally. Many famous artists from New York and abroad visit Detroit and Chicago, if not Kalamazoo.

And the people who live in Kalamazoo visit New York as well as Chicago and Detroit whenever they care to. New York is only fourteen hours away by turnpike, and less than three hours by plane. Travel is easy and natural in 1964, and it is not limited to those with high incomes.

In Krimsk travel isn't so easy. There is good bus service to Novorossisk and Krasnodar, but trips any farther away must be planned and scheduled well in advance. Almost no one has a car of his own, and the increasing number who do have motorcycles with sidecars don't find them convenient for long trips. To drive by motorcycle to Moscow is possible but not practical; to go by plane is easier, but in any case such a trip requires official approvals that aren't always obtainable.

Like Kalamazoo, Krimsk has begun to move from agriculture to industry; it is growing and most of its workers are employed in its factories, but it still has most of the way to go. A hundred years older than Kalamazoo in time, it is still thirty-five years or more behind Kalamazoo in its economic development. Since 1929, it has regained some of the time it lost as a drowsy Cossack village dominated by the feudal system of the tsars; it has begun to overtake Kalamazoo, but Kalamazoo is not standing still, and the economic gap between the two remains a wide one.

TWO

Trends Projected 1964 to 1999—
U.S.A. and U.S.S.R.

II

Economic Perspective

THE trends of economic growth stretch back across the centuries and forward to eternity. The last thirty-five years are a very small segment of time, but the lines that measure the directions of the change in even that small a segment can be projected forward or backward with some assurance that they are parts of the longer lines that will be a record of economic progress through the ages. Even the factors of war and depression had a relatively small total effect on the trends from 1929 to 1964. Those trends are part of the history of mankind, and are a product of many forces. The evolution of man includes his economic evolution and it is a continuing process.

The past is more than prologue to the future. Within the concept of time as a fourth dimension the past is part of the future. So it is that trends become significant and meaningful, for they are connected with all of time, before and after the years in which they have been noted and measured.

It is difficult, if not impossible, to measure trends exactly for even as short a period as thirty-five years. The figures are available, and perhaps they are true, but their meaning has changed. They haven't the same significance from year to year, and it is impossible to adjust them perfectly to a base year, or to translate them into exactly equal dollars and units of production. But if the lines are drawn as carefully and accurately as possible, it is evident that they can be extended backward into the past or forward into the future.

To project the economic trends in the U.S.A. (from 1929 to 1964) forward to 1999 is no more difficult and only slightly more impractical than to project them back to 1900. Their directions are apparent and it is unlikely that their rate of progress will be estimated with too much optimism. Although it won't be possible to check the accuracy of these projections until the thirty-five years are out, the chances are in favor of their being too conservative.

In any case, they must start from here, from 1964. What will be remembered of the U.S.A. in 1964? Increased racial tensions, new adventures in pursuit of the moon, nationwide labor disputes, a $20 million movie and a $200,000 golf tournament, the election of a new President, an improved technique for creating nuclear power—all are possibilities.

The great highways, turnpikes, and freeways stretching from coast to coast will be remembered, as will the facilities for traveling by air from any place in the U.S.A. to anywhere else in the world in less than twenty-four hours. The millions of cars and the traffic jams will be remembered, too.

In 1964 there are new houses and schools everywhere, office buildings and civic centers, new parks, stadiums, and airports. Only the railway depots and terminals are ancient and shabby. Hunger and despair are rarely seen on the highways or the public squares, although they still exist in the tenements and back streets.

The U.S.A. has an economy that is healthy but still apprehensive. The economic community fears that it is living on borrowed time as well as borrowed money, and that it may be getting better than it deserves. It is confident that it can maintain its growth rate and that it can avoid depressions and other economic drags, but it still has some lingering doubts. There are too many factors that aren't yet controlled, including the relationship of wage rates and hourly incomes to productivity and prices, the irresponsible demands of both labor and management, wasteful stock manipulation and speculation, and the impact of foreign production and foreign markets. We have invented the functions of industrial relations, industrial engineering, and motivation research, but we are unsure of the capacity of these new inventions to cope with the imminent problems of unemployment, industrial expansion, and international competition.

In 1964 the U.S.A. is on the verge of achieving a new level of economic competence and efficiency that will follow a scientific examination and definition of its problems.

It is encouraging to note that these problems are recognized and admitted by industry, but it is unfortunate that the specters of too much government or the threat of communism are invoked many times when objective analysis might produce more suitable and practical solutions. With greater objectivity and less instinctive fear of changes that may be overdue, the progress to be made in the next thirty-five years could far exceed any estimates based on mathematical projections of recent trends. Nuclear power and automation teamed with market research could produce results that can hardly be imagined today.

In the U.S.S.R. the directions of the economic trends are equally clear. Despite the setbacks that occurred during the war, the economy has moved steadily ahead along the same lines since 1929 or before. After 1942, it quickly regained its earlier momentum, and in 1964 both the pace and the direction of the trends are basic elements in the Soviet program for the future. The economic leaders in the U.S.S.R. think they know where they are going even if they don't know just how long it will take them to get there. It would be easy to estimate their progress with too much optimism based on the record since 1942, and it will be essential to check the accuracy of their projections regularly against what actually does happen.

What is being talked about in the U.S.S.R. in 1964 are the miles of nearly identical apartment buildings going up along the new wide boulevards, the internationally recognized successes of the Soviet astronauts, the deals for wheat with Canada and the U.S.A., the political rifts with China, the low average standard of living, and the barriers raised between Russia and the Western world.

The new Kremlin Congress Theater and the Bolshoi Ballet and vacations on the Black Sea will be remembered by Russians, as will the Metro in Moscow and the endless lines of people on street corners waiting their turn to buy grapes or chickens or tomato paste from the new crop in Krasnodar. In Soviet Russia in 1964 there are schools everywhere, high schools, nurseries, kindergartens, and institutes. Almost everyone is going to school part time; everyone is learning in night classes, seminars, factory meetings, and industry conventions.

Most people are poor, but they aren't hungry or cold. They have security and confidence in their future and the future of their State. Most believe in communism and think they know what it is. They are almost at the point where they don't need to argue about it as they used to. The Soviet economy is healthy in a modest way despite the shortage of grain in 1963. Industry continues to expand and there is always something to eat. There are white shirts with button-down collars, furs, practical raincoats, and plenty of good shoes and boots for sale, and the prices are about the same as they would be in Germany or the U.S.A. New recordings of

American jazz and Soviet folk songs are popular and inexpensive, and anyone can buy a motorcycle if he saves his money and waits his turn.

The trends in Soviet Russia are all sharp and steep. The number of homes with hot running water increases at a rate of 5 percent or 10 percent a year in many cities, new housing is being built in Moscow at a rate of 4 million square meters a year, enough for 400,000 people. The canning industry has increased its production 500 per cent since 1942. Whether these are realistic figures and whether they can be projected into the future is difficult to know.

What is not clear in 1964 is whether the economic problems facing the U.S.S.R. in the years ahead are known and admitted by the Soviet leaders and industrial managers. No one cares to confess that the projections to 1980 announced by Khrushchev are visionary and impractical. Consequently, it is impossible to find out what solid plans for growth are being made that can be achieved.

Whether the U.S.S.R. would survive at all was a legitimate question in 1929. The Revolution was only thirteen years old in 1930 and the atmosphere was desperate, elated, and stubborn. In thirty-five years the U.S.S.R. has grown up. It has survived famine and war. It is desperate no longer and its elation has moderated. It is still confident and stubborn, but it is beginning to see its potential in a longer perspective.

A projection forward to 1999 of the trends that have been observed and measured in the U.S.A. and the U.S.S.R. for the thirty-five years from 1929 to 1964 will at least put these trends into a perspective that will provide an indication of the relationship that may exist between these two economies if the present trends continue. Even purely mathematical projections would be useful for this purpose, but there is no reason why the mathematical projections cannot be modified to take into account pressures and changes that must be anticipated.

12

Economic Growth

THAT the gross national product of the U.S.A. will increase from $510 billion to $1,700 billion while that of the U.S.S.R. increases from $250 billion to $1,000 billion is not a prediction but only a conservative estimate of what could happen if there is peace for another thirty-five years. It may be too conservative. By 1999, the GNP of the U.S.A. may well pass the $2,000 billion mark, and the U.S.S.R. may not be far behind.

A difference of only 1 percent or 2 percent in the annual rate of growth could change these estimates completely. If the U.S.A. should go ahead at a rate of only 3 percent instead of 3.5 percent, and if the U.S.S.R. should grow at a rate of 5.5 percent instead of 4 percent, then the U.S.A. would reach a level of only $1,430 billion while the U.S.S.R. would attain a GNP of $1,630 billion.

A 3.5 percent annual rate of growth may be difficult for the U.S.A. to maintain. Soviet economists are confident that it will fall below 3 percent. And an estimate of only 4 percent for Soviet Russia would make the same economists laugh. They are sure that they will continue to increase their *national income* by 8 percent or more each year, and can't imagine any setbacks, barring nuclear war, that could bring the average growth of their GNP over the next thirty-five years down below a figure of 5 percent. Although the Soviets don't use the concept of *gross national product* in their own planning, they believe with some justification that if the Soviet GNP were measured, it would have increased at a rate of about 6 percent per year over the last decade.

Because the Soviet economy is entering a stage of development and growth that has been observed before in other places, it is relatively easy to predict that Soviet productivity will continue to increase at about its present rate. It may move ahead at a more rapid pace than that of the U.S.A. from 1929 to 1964, but since it is starting from a base similar to that of the U.S.A. in 1929, it is possible to anticipate some of the factors that may moderate its speed. An increase of only 400 percent in thirty-five years is a conservative guess, and the Soviet GNP will probably exceed the $1 trillion level before 1999, but not in 1980 as Khrushchev predicted.

In the U.S.A. the situation is different and less clear. The economy of the U.S.A. is at a level that has never been attained before and is in a stage of growth that is a completely new experience in the world. No one knows what can be accomplished, but it is logical and reasonable to project the current growth rate. A figure of $1.7 trillion in 1954 dollars, about $2 trillion in 1964 dollars, may even be too low.

It is true that Soviet economists believe that it will be impossible for the economy of the U.S.A. to continue to increase its total production by as much as 3.5 percent or 4 percent per year. For a few years it looked as though they might be right, but in 1963 and 1964 there is evidence that even a 4 percent rate is attainable. What remains to be tested is their reasoning concerning the pressures in a capitalist economy that inevitably cause cycles of growth and depression. Their thinking suffers from incomplete knowledge and understanding of the economy of the U.S.A. and from a tendency to use the word capitalism in an outdated Marxist sense.

Economists in the U.S.A. are misled, too, by the Communist label that it attached to the economy of Soviet Russia. They find it difficult to appraise the potentials for growth that exist in the industrial economy that has come into being in a communistic society. Their reasoning is conditioned by their fears and prejudices. They may be right in thinking that Communist values are inconsistent with maximum economic growth but they are talking politics and not economics when they reach such conclusions.

In the years ahead both the U.S.A. and the U.S.S.R. will have to answer many of the same questions. The progress of both economies will depend on their ability to cope with the problems of automation, employment, and marketing. Their GNP's will reflect the extent of their success in expanding their industries to provide work for all of their labor force, and to produce increasing quantities of the many things that are wanted in the world. Both will have to teach their workers the skills that are needed for automation, and learn how to use research to find out what additional goods can be produced and distributed at a profit.

The U.S.A. is already deeply involved in the problems of marketing and employment. Each year it is necessary to find markets for 3.5 percent more production in order to maintain a growth rate of 3.5 percent, and it is also necessary to find jobs for more than a million new workers. Meanwhile, the U.S.S.R. is trying to develop an industrial labor force adequate to maintain its rate of progress toward mechanization and automation. Its main concern is still to build and equip additional productive units and to find skilled labor to run them, but it is already beginning to consider the marketing problems that will exist when it achieves a somewhat higher level of total production.

It is apparent that the U.S.A. needs new markets for its production outside of its own domestic boundaries, and that it must find a way to expand its trade with the rest of the world. It is also clear that the U.S.A. has abundant opportunities to increase its foreign business, and that the U.S.S.R. could be one of its best customers. Many other countries need the tools and machines that the U.S.A. could supply, and it is reasonable to believe that the people of the U.S.A. would willingly accept in return the wines, spices, and perfumes and all the other useful and beautiful things that could be obtained from other parts of the world.

The possibilities of trade with the U.S.S.R. should not be put aside for political reasons that may be unsound. It is possible that trade with the U.S.S.R. might be as helpful to the U.S.A. as to Soviet Russia. It is unlikely that the Soviet industrial growth rate of 8 percent would be increased by trade with the U.S.A., and the U.S.A. needs additional foreign trade to maintain its own rate of growth. It is possible that the chances of the U.S.S.R.'s overtaking the U.S.A. in total production of any kind, national income or GNP, would be reduced rather than increased by such trade.

The deal for wheat with the U.S.S.R in 1963 and 1964 is a special case, but it is an example of a probable benefit to the U.S.A. without a corresponding advantage to the U.S.S.R. in terms of economic growth. A shortage of grain due to a poor crop forced the U.S.S.R. to use gold to buy grain instead of machines, and to a degree the purchase slowed the expansion of the Soviet economy.

In conversations with economists and plant managers in Moscow, Kharkov, and Krimsk it was apparent to me that industrialists in the U.S.S.R. would like to trade with the U.S.A. There were many new machines to be seen in the

Soviet factories, mostly from the satellite countries, but the people I talked to asked repeatedly about the kind of equipment that might be obtained from the U.S.A. They were aware of American technical progress in general, and often knew of specific items that could be useful to them. It was also made clear that trade with France and Spain and other areas outside the Iron Curtain was increasing, and would continue to increase.

Soviet Russia's growth may be delayed by the unwillingness of the U.S.A. to supply some of the tools she needs, but the expansion of our industry may be delayed, too, by our failure to produce all of the things we could have produced. Nor can we prevent the U.S.S.R. from getting adequate supplies of capital goods elsewhere. All that we can do is to make her growth more difficult and direct her trade to other areas that may eventually share a part of it with us.

Why we should want to hold back the economic growth of the U.S.S.R. is a good question. It would be difficult to argue that the dangers of communism will be increased as the U.S.S.R. achieves higher standards of living and greater economic stability. It is even possible that the proponents of communism are encouraged by the evidence of fear and antagonism that they can find in the trade barriers imposed by the capitalist nations.

It is also difficult to contend that the best way to win the economic race with Soviet Russia is to put as many obstacles as possible in her path. Sportsmanship aside, few top performers in any field have ever succeeded by holding back or harassing their opponents. The American way has always been to concentrate on developing the skills and strengths required to be best. There are two good reasons why the same rules should apply to the economic rivalry between the U.S.A. and the U.S.S.R. For one thing, policies and activities that are aimed at holding back one economy may very well hold back both. Furthermore, if both economies do move ahead as fast as they possibly can, they will still be unable jointly to keep up with the economic needs of the world. Politically, as well as economically, it would seem that expanded trade between the U.S.A. and the U.S.S.R. would be advantageous to both countries.

It is hard to understand how the prospects for communism will be improved when both the U.S.A. and the U.S.S.R. have economies producing more than $1.5 trillion of goods and services each year, and have average per capita incomes of $4,000 or $5,000. It seems more likely that the chances of war and the extension of communism will both be decreased when these goals are achieved. Other parts of the world will be able to see what has been accomplished with capitalism in the U.S.A. and with communism in the U.S.S.R. They will be able to measure the results objectively, and it is likely that most of the economic factors will favor capitalism. Only if the rate of growth in the U.S.A. did slow down would there be any justification on economic grounds for choosing communism.

The chances that the U.S.A. will make the kind of economic progress that has been described and will achieve the levels of growth that have been projected to 1999 are not just a matter of trends. The mathematical chances are much improved by the factors of enthusiasm and confidence. In the U.S.A. there is the necessary motivation to keep the economy moving dynamically ahead.

Everyone who thinks about it is enthusiastic about the prospect of continued economic growth. Union leaders, managers, stockholders, and investors all seek continued growth and larger incomes, and the rewards of expansion and growth are felt almost immediately in every sector of the economy. Housewives and secretaries, ballet dancers and movie queens, professional athletes and congressmen, mail clerks and scientists, whether they realize it or not, will share in a GNP that is measured in trillions because they all can expect to enjoy personal gains as a result of it.

The motivations are almost universal in the U.S.A. and they build up quickly into effective pressures to overcome any possible hindrances to continued growth such as railroad strikes, excessive profits and prices, restrictive taxation, and deflation. All the pressures on the economy reflect a nearly unanimous attitude that the economy should "go for broke" to achieve the maximum gains possible. As long as this is true, the chances are good that the projections made here will be exceeded even before 1999.

There are equally powerful motivations for growth in the U.S.S.R. The Soviet economy has quadrupled since 1929 and it is evident to anyone who visits the U.S.S.R. that expansion and change are still continuing at a rapid pace.

In any case it would be unrealistic to make plans based on the odd chances of catastrophes. An increase in the GNP of more than 300 percent is a modest projection of the present rate of growth in the U.S.A. It has not been modified to take into account the maximum potentials of automation and the possibility that new industries may be developed to make full use of the available work force.

If the growth rate increased from 3.5 percent to 4 percent, the GNP in 1999 would exceed $2 trillion, or $6,700 per capita. An increase to 5 percent would result in a GNP of $2.8 trillion, or $9,300 per capita, all in constant, comparable noninflated 1954 dollars.

Moreover, any slowing down in the current rate of growth would be a serious blow to the economy, and many pressures are already being exerted to push it even higher.

These are conservative estimates, but they are not pessimistic. They assume that there will be an opportunity for continued peaceful growth, that labor and management will learn to avoid wasteful delays and detours, and that additional unemployment will not offset increased productivity. They are reasonably modest, but they should be suitable for comparison with the prospects for the U.S.S.R.

If present trends continue it is estimated that total production in the U.S.S.R. in 1999 will reach a level of $1,000 billion annually. If so, production will be four times as large as it is in 1964, and it will be remembered that production in 1964 is nearly four times what it was in 1929. However, this projection assumes that the annual growth will come down during the next thirty-five years from the present rate of about 6 percent per year to an average of only 4 percent. With an increase of 6 percent each year the GNP would reach the trillion-dollar level in less than twenty-four years. Although it is possible that this rate can be maintained, it will become more and more difficult to achieve the same percent of increase every year as the economy continues to grow. Moreover, the current rate of 8 percent for industrial production will probably come down somewhat as the mechanization of industry becomes more complete.

If a growth rate of 6 percent could be maintained, then the Soviet GNP would be $1.9 trillion in 1999, and would probably have exceeded that of the U.S.A.

The figure for the U.S.S.R. of $1,000 billion is about 60 percent of the projected GNP of the U.S.A. in 1999, and it should be noted that total production in the U.S.S.R. will consist of proportionately less consumer goods and services and more capital goods and governmental expenditures for defense and research. It is estimated that goods for household consumption will increase from $125 billion to $625 billion, which is about 42 percent of the U.S. figure. Production of capital goods will increase, according to this estimate, from $68 billion to $200 billion but the percentage of total production will come down from 27 percent to 20 percent. Total investments in capital goods will probably be somewhat smaller than in the U.S.A. where the total might continue to be as high as 15 percent of the GNP, or $250 billion. Expenditures by the government for defense, research, administration, and communal services would probably increase from $57 billion to $175 billion, reflecting in part some expenditures by the government for services that would be included in household consumption in the U.S.A.

In theory it is relatively easy to maintain a desired rate of growth in a totalitarian economy. The central planning authority can decide what additional production is wanted, and can allocate supplies, manpower, and capital to the sectors of the economy where they are required. It can place what amounts to a purchase order for the additional goods wanted, and can determine what these goods should be without considering the theories of supply and demand that concern a free market. The central planning authority can decide what additional consumer goods are to be produced, what further investments must be made in enlarged productive facilities, what expenditures there should be for research and development, and what amounts will be used to build schools and roads.

It can place orders for the additional production required to assure a growth rate of 6 percent in the GNP. It can even hedge against a 10 percent failure in reaching production goals by ordering 10 percent more than it expects to be delivered! It can then allocate the necessary manpower to the projects required to produce the additional goods with confidence that the manpower allocated will work as many hours as it takes to produce the goods that are required. The wages paid to the workers who produce the goods will be equal in total (after taxes that may be included in the prices set) to the value of the portion of the total production that consists of consumer goods that the workers can buy.

Many of the problems of a free economy are avoided. The production desired can be achieved whether or not workers are satisfied with their wages, and whether or not entrepreneurs are pleased with their profits. Any incentives or pressures needed to obtain delivery of the goods ordered may be invoked, and the chances are fair that the desired rate of growth will be achieved.

The chances are limited, however, by several factors that cannot be solved with legislation or force. One is the availability of labor to do what needs to be done. Not even the Soviet labor force can be expanded at will, and it takes time, patience, and imagination to develop the trained men needed to increase the rates of productivity per man hour, and to move toward automation. It is unlikely that the labor force in total can be expanded in the near future. Ninety-five percent of the men over sixteen years of age and 62 percent of the women are now employed, and the birth rate over the last fifteen years offers no hope for any increase in the number of people joining the work force each year. However, more than 40 percent of all workers are now employed in agriculture, and it is likely that many of these will find their way into industry as continued progress is made in mechanizing agriculture, and in increasing the agricultural yields.

Another problem is that of maintaining a high rate of investment in new factories at the expense of consumer goods and an improved standard of living. No one can tell how long the people of the U.S.S.R. will accept a standard of living less than 40 percent of that in the U.S.A. when they believe that their total production is rapidly overtaking that of the U.S.A. What they could do about it is not certain, but it would be unwise to assume that they will not find a way to exert pressure for improved living standards when it becomes apparent that improvements could easily be made. A shift toward greater production of consumer goods does not of necessity mean a reduction in the rate of growth, but it does complicate the problems of allocating resources and manpower, and enlarges the job of training men for a variety of new techniques and procedures. Such a shift would probably slow down the rate of increase in total production, although it would open new opportunities for Soviet industry to take advantage of technologies that have been brought to high levels of efficiency elsewhere in the world.

Another factor that might slow down the rate of growth is the increasing cost of depreciation. As the total capital investment in the U.S.S.R. continues to grow, the burden of maintaining that investment increases, and the cost of holding current levels of production becomes an ever larger percentage of the total budget for investment.

Most important of all is the problem of bureaucratic organization and red tape. The Soviet industrial organization is still noted for its administrative and operational inefficiency. Too frequently the lack of a key component will shut down a large factory for hours or days. Or unrealistic and impractical goals will result in the production of items that can most easily be produced in large numbers or large tonnage even though these are not the sizes and shapes needed by the economy. That a totalitarian economy can ever function as well as a market economy has yet to be demonstrated.

Nevertheless, all limitations considered, the chances are that the GNP of the U.S.S.R. will continue to grow at an annual rate somewhere between 4 percent and 6 percent. This is a conservative estimate, one that will not please the economists I met in Soviet Russia, but one that takes into account the many problems that must be solved and the obstacles that must be surmounted before the Soviet economy can honestly hope to compete with that of the U.S.A. It is quite possible that these estimates will be exceeded. I, for one, hope so, and think that it will be to our advantage, as well as theirs, if they are. In any case, a GNP of $1,000 billion is a practical probability.

How this total production may be divided between consumer goods, capital investments, and governmental expenditures for other purposes will depend on the needs of the Soviet economy over the next thirty-five years. In addition to what would be produced at the present level of $250 billion per year, it is estimated that an increment of at least $13 trillion will have been produced by 1999, an average increase of $375 billion per year. How this total increment is divided between capital investment and consumer products will determine what kind of additional goods will be desired in 1999.

At present the needs of the Soviet economy are probably rated in the U.S.S.R. according to a scale something like this:

1. Military needs, items for defense.
2. Needs for nuclear and space research.
3. Capital investments needed to create an industrial economy as large as that of the U.S.A.
4. Basic commodities for consumers, including agriculture.
5. Education, other public services.
6. Household goods, appliances.
7. Recreation, travel.
8. Automobiles, housing, roads.

Military needs come first, and it is likely that an additional amount of as much as $2 trillion will be spent in total by the government for defense and research over the next thirty-five years. But this still leaves more than $11 trillion for capital investment and consumer goods. How much will be needed to complete and maintain an investment in factories and machines adequate to produce all of the goods, machines, tools, food, appliances, and housing required for a population of 350 million with a standard of living comparable to that of the U.S.A.? This is a calculation that cannot even be attempted, but it is beside the point, anyway, for it assumes that the objective of the U.S.S.R. will be to create an economy that can give all the people a standard of living like that in the U.S.A. It is more likely that the U.S.S.R. will be satisfied with a much lower standard of living and will choose to produce more capital goods for investment in satellite countries and in underdeveloped areas of the world. All that can be done is to note where the present trends are leading, and then to find out what those trends would produce.

Perhaps 25 percent of the total additional production in the U.S.S.R. over the next thirty-five years will consist of capital goods. If so, the total capital investment made by 1999 will nearly double what would be invested at the 1964 rate, and the annual rate of investment will increase from $68 billion in 1964 to $200 billion in 1999. In total, about $4.7 trillion will go into capital goods.

This figure compares with an estimate of $1.2 trillion for the period from 1929 to 1964.

In the U.S.A. the composition of the GNP is determined by the return on investment that can be earned with various items that might be produced. In a market economy the tendency is to produce what will be most profitable, and the assortment of goods produced is presumably the most profitable combination that could be put together.

In the U.S.S.R. the composition of the GNP should be what the central authorities judge to be most essential to achieving their long-range goals, regardless of profitability. Moreover, in the U.S.S.R. it is difficult, if not impossible, to determine whether an enterprise or industry is profitable or not. With the costs and allocations of raw materials arbitrarily determined, and with wage rates and prices on consumer goods arbitrarily adjusted, the efficiency of the various areas of the economy cannot be measured in terms of profits. In fact, it is hard to know how to measure the efficiency of the Soviet economy, or to compare it with any other economic system.

Consequently, the pressures that determine what will be produced are not related closely to profitability or efficiency.

They include basic needs that aren't satisfied for food, clothing, and minimum housing, needs for defense and international security, desired public facilities, schools, subways, and roads, requirements for enlarging industrial capacity and increasing total production, requirements for foreign aid and propaganda, and the desires of some critical segments of the population for more comfortable standards of living.

Having in mind these pressures, it is understandable that the GNP should be divided as it is in 1964, with about half of the total going into consumer goods, and 28 percent into gross investments.[1] No change in this pattern is likely to come about through increased demands for consumer goods. Living standards are improving at an accelerating rate, and in the last thirty-five years the U.S.S.R. has demonstrated the ability of a totalitarian economy to control the wishes of its people for more rapid improvements.[2]

In 1957 the U.S.A. put 66 percent of its GNP into goods for household consumption, and 19 percent into gross investment. The Soviet put a 50 percent larger share of its production into investment, and the U.S.A. put a 33 percent larger share into consumer goods.

In part this difference reflects a difference in the stages of growth of the two economies. The share going into household consumption increased in the U.S.S.R. from 45.5 percent in 1950 to 49.2 percent in 1955, and it is still increasing. However, the share going into investment remained stable at 27.9 percent during the same span of five years.

The present composition of the GNP could possibly continue for another thirty-five years. A continued investment of 28 percent would produce more than $6 trillion of additional productive capacity (less depreciation), possibly enough to

[1] Abram Bergson, *The Real National Income of Soviet Russia Since 1928* (Cambridge: Harvard University Press, 1961), p. 237.
[2] Robert W. Campbell, *Soviet Economic Power* (Boston: Houghton Mifflin, 1960), p. 161.

overtake the U.S.A. by the end of the century. But the chances are greatly in favor of a gradual increase in the percentage of the GNP going into consumer goods, and a corresponding decline in the percentage, not the total amount, going into investment. For one thing, if this adjustment were not made gradually over the next thirty-five years, the U.S.S.R. would find itself in 1999 with all the productive capacity needed for a standard of living approaching that of the U.S.A., and with a pattern of consumption substantially lower. It would have a surplus of goods that it could market or invest elsewhere in the world, and an undeveloped domestic market with many unsatisfied needs. If the U.S.S.R. does reach a GNP of $1,000 billion, it will then become clear whether the people of the U.S.S.R. will ever have a standard of living like that in the U.S.A., or whether a portion of the total production of the U.S.S.R. will regularly be set aside to assist in the development of other Communist countries.

The objectives of the Soviet economy in 1964 appear to be to achieve a stable, but relatively modest standard of living for all the Russian people, and to maintain a rate of growth that will enable the U.S.S.R. to gain a dominant position in many world markets. By eliminating a few luxuries, and by controlling wasteful extremes in packaging and product variety, the U.S.S.R. may achieve a standard of living nearly equal to that of the U.S.A. at 85 percent or 90 percent of the cost, and may thus have a larger share left to invest in world markets than is available to the U.S.A. Whether this difference will be an advantage or a liability will depend upon the point of view of the rest of the world. Those who have luxuries to sell may prefer to deal with the U.S.A. If both the U.S.A. and the U.S.S.R. did reach the same levels of production in 1999, it is likely that the Soviet Union would allocate less production for its domestic requirements, and more for foreign aid and investments.

If the present trends continue, the U.S.A. and the U.S.S.R.

combined will produce $34 trillion more during the period from 1964 to 1999 than they would have produced if they had maintained the 1964 levels of production.

This amounts to an average yearly increment larger than their *combined* GNP in 1964. Their combined total production in 1999 will approach a level of $3 trillion annually (1954 dollars). Of the additional goods produced, 38 percent will be made in Soviet Russia. These figures compare with increments of production from 1929 to 1964 of $4,430 billion in the U.S.A. and $3,237 billion in the U.S.S.R. or a combined total of about $7,667 billion. The additional goods to be produced in the next thirty-five years will amount to more than four times the total additional goods produced in the last thirty-five years. These increments amount to $69,400 per capita in the U.S.A. and $37,500 per capita in the U.S.S.R.

Because the U.S.S.R. will allocate a larger share of its total production for machines and other capital goods, its production of goods for household consumption from 1964 to 1999 will probably amount to only $25,000 for every person living in the U.S.S.R. in 1999. In the U.S.A. the total additional goods produced for personal consumption will amount to more than $50,000 per capita. Per capita consumption in the U.S.A. will still be 2.3 times that in the U.S.S.R. in 1999, but it will be down from 3.3 times as much in 1964.

If these figures seem to promise a level of affluence for nearly everyone, this is just about what they mean. During the next thirty-five years, the average family in both the U.S.A. and the U.S.S.R. will be able to acquire three times as many additional goods for household consumption as a similar family could afford to buy from 1929 to 1964. Soviet living standards will actually increase at a rate faster than living standards in the U.S.A. have increased since 1933. There will be an abundance of comfort in both countries, with very few families left at levels of want, and there will be comfortable abundance for more than half of the families

TABLE VII

ECONOMIC GROWTH FROM 1929 TO 1964, PROJECTED TO 1999

(figures are constant 1954 dollars)

U. S. A.

	1929	1964	1999
Population (million)	122	190	300
Labor Force (million)	49	75	120
Employed (million)	47.6	70.5	114
GNP (billion)	$181.8	$510	$1,700
Personal Income (billion)	$139.2	$420	$1,500
Federal Budget (billion)	$5.9	$77	$170
Personal Consumption (billion)	$128.2	$344	$1,230
Per Capita GNP	$1,492	$2,680	$5,666
Per Capita Income	$1,140	$2,210	$5,000
Per Capita Personal Consumption	$1,050	$1,810	$4,100
Hours Worked Per Year Per Worker	2,250	1,940	1,680
Total Hours Worked (billion)	107	136	191.5
GNP Per Hour Worked	$1.70	$3.75	$8.87

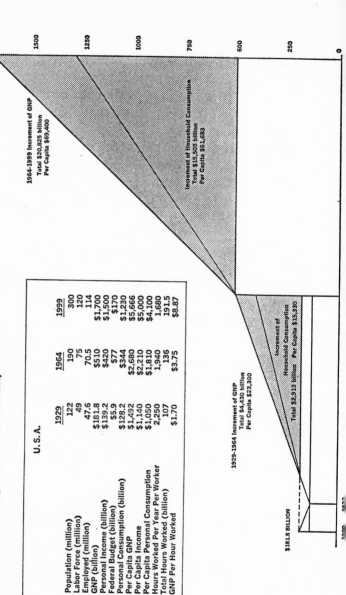

$1700 BILLION

1500

1250

1000

750

500

250

0

1964-1999 Increment of GNP
Total $20,825 billion
Per Capita $69,400

Increment of Household Consumption
Total $15,505 billion
Per Capita $51,683

1929-1964 Increment of GNP
Total $4,430 billion
Per Capita $23,300

Increment of
Household Consumption
Total $2,913 billion   Per Capita $15,330

$181.8 BILLION

ECONOMIC GROWTH FROM 1929 TO 1964, PROJECTED TO 1999
(figures are constant 1954 dollars)

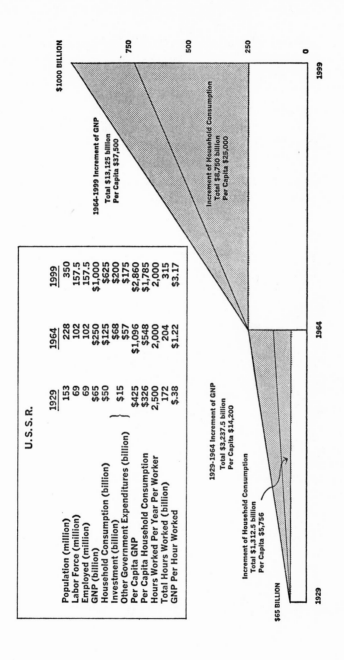

U.S.S.R.

	1929	1964	1999
Population (million)	153	228	350
Labor Force (million)	69	102	157.5
Employed (million)	69	102	157.5
GNP (billion)	$65	$250	$1,000
Household Consumption (billion)	$50	$125	$625
Investment (billion)	$15	$68	$200
Other Government Expenditures (billion)	}	$57	$175
Per Capita GNP	$425	$1,096	$2,860
Per Capita Household Consumption	$326	$548	$1,785
Hours Worked Per Year Per Worker	2,500	2,000	2,000
Total Hours Worked (billion)	172	204	315
GNP Per Hour Worked	$.38	$1.22	$3.17

1964-1999 Increment of GNP
Total $13,125 billion
Per Capita $37,500

Increment of Household Consumption
Total $8,750 billion
Per Capita $25,000

1929-1964 Increment of GNP
Total $3,237.5 billion
Per Capita $14,200

Increment of Household Consumption
Total $1,312.5 billion
Per Capita $5,756

$65 BILLION

in the United States. In the U.S.S.R., the level of abundance will probably still be limited to a relatively few industrial and political leaders. It is unlikely that the percentage of affluence in the U.S.S.R. will even approach that of the U.S.A. in 1964.

Although it should be emphasized that all of these projections and estimates are subject to large factors of error, Tables VII and VIII illustrate what has happened since 1929, and indicate what can happen in the future if these trends continue along their present lines. These tables include most of the figures that have been quoted already for the period from 1929 to 1964, and also show comparable estimates for the economies of the U.S.S.R. and the U.S.A. in 1999.

13

Incomes and Living Standards

A VISIT to Soviet Russia will convince a thoughtful observer that living standards can't be measured solely in terms of per capita incomes. The way of living and the meaning of the various components of the standard of living must be weighed in comparing living standards in the U.S.S.R. and the U.S.A. It is possible that a Soviet citizen would rate a communal swimming pool higher than a private swimming pool in making such a comparison. He may even feel sorry for the poor Americans who have to drive miles to work in their own cars at great expense to themselves in time as well as money while he can get to work in a few minutes on the Metro. He may even believe that having an apartment like everyone else is a higher level of living than having separate private houses that encourage selfish private thinking. Certainly to the extent that he willingly foregoes the seeming advantages of private housing, he does eliminate many costs that must be assumed by Americans and that could be deducted from their nominal incomes in making a comparison of their living standards with those of similar income groups in the U.S.S.R.

It is certain that the Soviet wage earner does have a basis for arguing that his lower dollar income is equivalent to much higher incomes in the U.S.A. He not only saves many expenses, for lawn mowers, garden tools, garages, furniture, lamps, china, and silver, but he also saves hours of time that he would have to spend in maintaining and furnishing a home. On summer evenings and on weekends he has time for other activities that he may feel add more interest and significance to his life than would the same amount of time spent in cutting the lawn and fixing the screens.

He has no room in his apartment for more than one or two visitors at a time, and he doesn't need the variety and quantity and quality of equipment, furnishings, and tableware that are found in private homes elsewhere in the world. He doesn't believe in private parties, anyway. He sees his friends in meetings and seminars and official celebrations; and in a society in which the individual is not important, it is safer for him to remember his wedding anniversary with a night out for dinner and dancing at the Metropole or the Nacional.

Although per capita income in the U.S.S.R. will continue to lag far behind that in the U.S.A., it is likely that the improvements in Soviet living standards will be greater and much more important to the average family in the U.S.S.R. than the gains expected in the U.S.A. The most significant change promised in the U.S.A. will be the almost complete elimination of the income groups with less than $1,000 per capita per year. If present trends continue, it is estimated that the average per capita income of the lowest two-fifths of the population will be $2,250 in 1999. This is the equivalent of more than $7,500 per family. Perhaps 2 percent of the people will have incomes as low as $1,200 per capita.

A minimum per capita income in the U.S.A. of $1,200

will compare with a minimum of about $1,000 in the U.S.S.R., and the average income of the top 20 percent of the population in the U.S.A. of about $9,750 will compare with an average income for the top 3 percent in the U.S.S.R. of $5,400.

The trends point to an increase in living standards in the U.S.A. of more than 100 percent by 1999. Average per capita income of $5,000 will reflect the increased productivity of the American economy. This would be an increase from approximately $2,210 (in constant dollars) in 1964. The actual numbers may be much higher if inflation continues along its present course.

What is more difficult to estimate is what the increase will mean. How far will it go in extending a degree of affluence to the lower income groups of 1964? How much of the increase will flow into vacations in Italy, country clubs, modern art, and hi-fi equipment, and how much into improved plumbing, low-cost housing, and better education?

In 1964 it is estimated that 76 million Americans are still living in poverty or deprivation with family incomes of $4,500 or less. Of these half have family incomes under $3,000. Although this estimate of 76 million is down from 96 million in 1929, a reduction of 21 percent, it is evident that there is room for improvement. A similar gain in the next thirty-five years might reduce the number of deprived and poverty-stricken to 60 million which would still be 20 percent of the population.

What are the trends? Is an improvement of better than 20 percent a practical possibility in the next thirty-five years?

If the lower two-fifths of the population receive only 2 percent more of the total personal income than they received in 1964, and if total income increases at the same time by nearly 30 percent, the result is that the average per capita income of the lowest two-fifths is increased from $880 in 1964 to $2,250 in 1999.

TABLE VIII

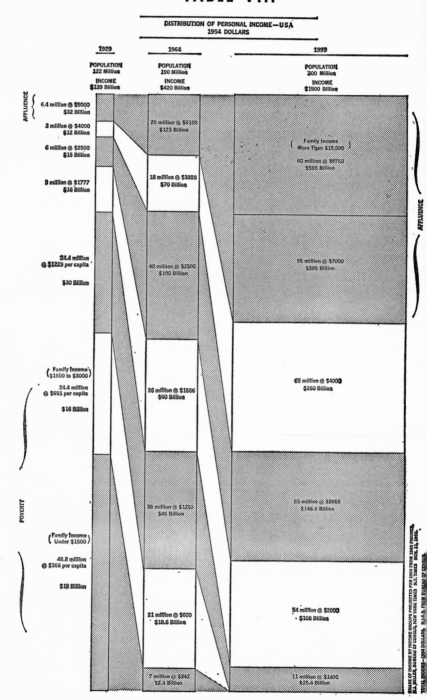

DISTRIBUTION OF PERSONAL INCOME—USA
1954 DOLLARS

1929

POPULATION
122 Million

INCOME
$139 Billion

1964

POPULATION
190 Million

INCOME
$420 Billion

1999

POPULATION
300 Million

INCOME
$1500 Billion

AFFLUENCE

6.4 million @ $5000
$32 Billion

3 million @ $4000
$12 Billion

6 million @ $2500
$15 Billion

9 million @ $1777
$16 Billion

24.4 million
@ $1229 per capita

$30 Billion

20 million @ $6150
$123 Billion

18 million @ $3888
$70 Billion

40 million @ $2500
$100 Billion

(Family Income)
More Than $15,000

60 million @ $9750
$585 Billion

55 million @ $7000
$385 Billion

AFFLUENCE

(Family Income)
$1500 to $3000

24.4 million
@ $655 per capita

$16 Billion

36 million @ $1666
$60 Billion

65 million @ $4000
$260 Billion

POVERTY

38 million @ $1210
$46 Billion

55 million @ $2665
$146.6 Billion

(Family Income)
Under $1500

48.8 million
@ $368 per capita

$18 Billion

31 million @ $600
$18.6 Billion

54 million @ $2000
$108 Billion

7 million @ $342
$2.4 Billion

11 million @ $1400
$15.4 Billion

SHARE OF INCOME BY INCOME GROUPS PROJECTED FOR 1964 FROM 1960 FIGURES,
H. MILLER, BUREAU OF CENSUS, NEW YORK TIMES NOV. 14, 1962.

TOTAL INCOME—1954 DOLLARS, N.P.A. FROM BUREAU OF CENSUS.

It is unlikely that any actual poverty will still be permitted in a $1.7 trillion economy, but there may be a few families with less than $100 per week. Hunger and serious deprivation will be eliminated in the U.S.A. by the end of this century.

If a family income of $6,000 is required for a decent standard of living, it is estimated that more than 85 percent of the population should have such a living in 1999. The lowest ten percent would have an average per capita income of $1,570, or more than $5,000 per family. (If the average family consists of 3.4 persons the per capita income would be $5,350.) Only a very few families would have less than $4,000 per year.

In 1999, the lower two-fifths would have an income of $270 billion, or $2,250 per capita, an increase of 256 percent over 1964. Only a small percent, less than 5 percent, would have incomes as low as any of those in the lowest two-fifths in 1964.

The middle 40 percent, 120 million strong, will have average per capita incomes of $5,375. Nearly all will live in families with incomes in excess of $15,000.

Those in the top 20 percent will have average per capita incomes of $9,750—family incomes of more than $30,000 on an average.

Table VIII indicates about what will happen to various income levels in the next thirty-five years if the economy does reach a total GNP of $1,700 billion, if personal income reaches a total of $1,500 billion, and if the income is distributed only a little more equally than it is distributed in 1964. It is not a prediction of what will happen but an approximation of what must happen if the present trends continue.

Note that the percentage of families with incomes of $4,500 or less is down from 80 percent in 1929 to 40 percent in 1964, and will be only 4 percent in 1999. Meanwhile the families with incomes in excess of $15,000 have increased from 5 percent in 1929 to 10 percent in 1964, and will include 38 percent of all families in 1999. Almost 10 percent will have family incomes of more than $30,000. The percentage of families with incomes ranging from $4,500 to $15,000 is up from 15 percent in 1929 to 50 percent in 1964, and will go to 58 percent in 1999.

It has already been estimated that in the U.S.S.R. the production of goods for household consumption will increase from $125 billion in 1964 to $625 billion in 1999. This would be an increase of 325 percent per capita, from $548 to $1,785.

If these estimates are realized, in 1999 the Soviet standard of living will still be less than half as high as that of the U.S.A., about equal to 1964 living standards in the U.S.A. when allowance is made for communal benefits. The U.S.S.R. will have total production equal to 60 percent of that in the U.S.A. but will still be thirty-five years behind in her standard of living.

However, these mathematical projections, even if they turn out to be approximately correct, don't tell the whole story. It is unlikely that Soviet families will ever have housing like that in the U.S.A., nor will as many families own even one automobile. The cost of housing and transportation will take a relatively small share of the budget for household consumption, and thus will leave a larger share for other consumer products and services. Many other costs are much lower in the U.S.S.R. than in the U.S.A. Medical and hospital services, education, and insurance are paid for by the State, and their cost is only partly included in figures for household consumption.

If the worker in the U.S.S.R. who has an average income of $1,785 has to spend only 10 percent of that income for housing, education, insurance, and medical expenses, and has $1,607 left for other purchases, he may be nearly as well off as the average person in the U.S.A. in 1964 who has an income of $2,210, spends 30 percent for these same items, and has only $1,547 left for other expenditures. He won't actually be as well off because he won't have a comfortable home of his own and a private car to take him wherever he wants to go. But he won't have to invest in lawn mowers and furniture and gasoline, and may have more to spend on theaters and music and travel. The chances are that in many ways he will have reached a level of affluence in 1999 approaching that of the average American in 1964. The chances are good, too, that he will have begun to feel a need for better housing and transportation, and will have made a start in the direction of satisfying that need.

The meaning of a per capita income figure depends on how income is actually distributed by income levels. Per capita income of $1,785 in 1999 may or may not be enough to give everyone a comfortable standard of living. If a minimum family income of $5,000 is enough for comfort (per capita of $1,470), then the top 10 percent of the population could have family incomes of nearly $12,000 and per capita incomes of $3,500, and the middle 50 percent could have average incomes of $1,700 per capita, or $5,570 per average family. (This assumes an average family of three persons, which may be low.) In this case, incomes would be distributed about like this:

U.S.S.R.—1999

35.0 million @	$3,500 =	$122.5 billion
175.0 million @	1,700 =	297.5 billion
140.0 million @	1,400 =	205.0 billion

With such a distribution of income, there would be very few families in the U.S.S.R. with incomes as low as some families in the U.S.A. today. Statistically, the U.S.S.R. would have adequate living standards for even the lowest income groups. In fact, it is certainly likely that there will be some backward areas in the U.S.S.R. like Appalachia in the U.S.A. of 1964, where conditions of primitive poverty still exist.

It is unlikely that a perfect socialistic distribution of income will be possible. At least a few top officials, political and industrial, will probably enjoy family incomes of at least $15,000. Many directors and specialists are now paid as much as $10,000 or $800 per month. Other families with wage earners contributing $60 to $1,800 per year may have to settle for average per capita incomes of no more than $1,000.

A more realistic distribution of income in 1999 would probably look like the data in Table IX.

TABLE IX

DISTRIBUTION OF INCOME—U.S.S.R.—1999

POPULATION (in millions)	PER CAPITA	PER FAMILY	TOTAL (in billions)
10	$5,400	$16,200	$ 54.0
25	3,500	10,500	87.5
175	1,700	5,100	297.5
100	1,460	4,380	146.0
40	1,000	3,000	40.0
350	1,785	$ 5,355	$625.0

It must be emphasized that income in the U.S.S.R. is not measured solely in terms of wages or other payments in cash that are available for the purchase of any consumer goods or services desired. It includes expenditures made directly by the State for items of household consumption that are allocated arbitrarily to individuals and families. The figures released for total household consumption are used here, and per capita income is presumed to be a per capita share of total household consumption. These figures do include some but not all of the costs of education and other communal services.

No matter how income is distributed, the present trends point to living standards in 1999 that will compare favorably with those in the U.S.A. in 1964. Most of the population will live in relative comfort, and many (10 percent) will be as well off as the upper middle group in the U.S.A. in 1964. A few may even be affluent. None will live in actual poverty.

Relatively few political and industrial leaders may have an unlimited supply of all of the things that they want and more services than they need. Like their opposite numbers anywhere else in the world, they will have first choice of the available housing, transportation facilities, art objects, and theater tickets.

Only slightly less well off will be a fairly large class of technicians, scientists, administrators, managers, generals, and controllers who will also have nearly anything they need and will only have second choice of privileges in short supply. The one thing they won't have that is popular with the same class elsewhere will be an opportunity to speculate on the growth of their economy. They won't have stocks and bonds. Nor will they be able to acquire a vested share in the total production of the U.S.S.R. (Possibly they will find a way to invest their surplus income in capitalist stocks.) There are no indications in present trends that any changes are contemplated in the basic Communist rules with regard to private property and interest rates.

In 1964 more than half of the people in the U.S.A. live in families with incomes of $5,000 or more, and in 1999 more than half of the people in the U.S.S.R. will have attained a similar level. More than half will live in families whose total income exceeds $5,000 and who may enjoy living standards that are better in many ways, excluding housing, than those of families with the same incomes in the U.S.A.

The rest of the population will still have adequate incomes of at least $3,000 per family, or possibly the equivalent of more than $4,000, after taking into account the communal benefits supplied by the State. There will be no poverty, yet nearly everyone will be relatively poor by the standards of the U.S.A.

Whether top incomes of $8,000 to $10,000 per family will be sufficient to spur and reward the millions of minds that will be needed to maintain the rate of growth required to reach a GNP of a trillion dollars or more by 1999 remains to be seen. The argument of Soviet economists would be that the growth rate will be maintained by sensible planning that is not related to personal profits, and that the necessary brains will be available at a fair price.

However, it is unlikely that any great changes will be made in the distribution of income in the next thirty-five years. There will continue to be a wide spread between the earnings of the ordinary farm laborers and unskilled workers and those of the managers and specialists who direct the growth of the economy.

It was estimated that incomes in the U.S.S.R. varied by at least 600 percent in 1964. In 1999 the difference in family incomes may be reduced somewhat, but the difference in wage rates may actually increase from a minimum wage scale of $50 per month to a high of $1,000 or more.

In 1964, the top 5 percent of the population had incomes at least six times as large as the average incomes of the lowest 40 percent, and in 1999, it is estimated that the top 5 percent will only have a little more than four times the average of the lowest 40 percent. As in the U.S.A. the incomes of the lowest two-fifths will increase somewhat faster than the incomes of those in the middle and upper groups.

According to these estimates, the lowest two-fifths will get 30 percent of the total personal income in 1999, an increase from 24 percent in 1964. Their per capita income will increase by four times.

The incomes of the middle 50 percent will increase nearly three times, and those of the top 5 percent will more than double (see Table X).

TABLE X

INCOME LEVELS—U.S.S.R.

| | 1964 | | | 1999 | |
LEVEL	POPULATION (in millions)	PER CAPITA	LEVEL	POPULATION (in millions)	PER CAPITA
Top 5%	11	$2,000	Top 10%	35	$4,000
Middle 55%	126	576	Middle 50%	175	1,700
Lowest 40%	91	334	Lowest 40%	140	1,330

Although there will still be a wide difference in income levels in 1999, it could be contended that the progress made toward ultimate Communist goals will have been entirely satisfactory. "To each according to his needs" does not promise a completely equal distribution of income. Some may have greater or different needs than others in any group of people, and even in a Communist society it may be necessary to recognize that some of those who work harder do so because they feel that their greater needs should be satisfied. The difference between communism and capitalism may be found in their definition of needs.

14

Population and Employment

BY 1999 it is estimated that there will be 300 million people in the U.S.A. and 350 million in the U.S.S.R.. The work force will increase from 75 million in 1964 to 120 million in 1999 in the U.S.A., and from 102 million to 157.5 million in the U.S.S.R.

The hours worked will increase by 39 percent in the U.S.A. to 191.5 billion and by 54 percent in the U.S.S.R. to 315 billion. Current trends indicate that unemployment in the U.S.A. will continue to increase (although numerical unemployment will remain at about 5 percent), and that as a result living standards will continue to be lower than they need be.

Trends of unemployment are not projected in the U.S.S.R. because it is more likely that unemployment can be circumvented in a command economy at the growth stage of the U.S.S.R. However, there has already been some decline in the hours worked per worker in the Soviet Union, and it is very possible that some further decline will be unavoidable.

Unemployment is a problem in both areas. In the U.S.S.R., the lack of machines and tools slows down the rate of employment. In the U.S.A., the lack of markets to take more goods at a profit is a more critical problem.

The problem of employing all the available labor in the most efficient way has not been solved in either the U.S.A. or the U.S.S.R. In the U.S.A., the result is unemployment. In the U.S.S.R., the result is inefficient employment. Both are costly and wasteful. Trade between the U.S.A. and the U.S.S.R. would help both economies. The U.S.S.R. needs capital goods that could be produced with the unemployed labor in the United States. In both economies the trend seems to be toward greater unemployment (less efficient employment) of a better trained and more efficient labor force.

The population of the U.S.A. will continue to increase during the next thirty-five years, and the rate of increase could even accelerate as a result of the increased security that is expected and the increase in total wealth. It is only a guess that the figure for 1999 will be 300 million, but it is a conservative guess based on a projection of the present direction with some allowance for the tendency toward earlier marriages and more children which may be offset by the demand of more families for comfortable standards of living. Whether the actual figure turns out to be 300 or 325 million makes little difference in the significance of this trend. What is important is that the population will continue to grow at a relatively rapid rate and that the increase in population will create new demands for goods and for jobs.

The tendency of the population to concentrate in the cities and in centers of industrial production is significant, too, for it runs parallel with the direction of industrial growth. It is likely that more people in total, and more as a percentage of the total population, will live in large urban complexes made up of groups or chains of cities filling entire regions of the country. This is a trend that is already under way. It is anticipated that the area from Boston to Washington, D.C., will comprise such a semiurban axis, and that similar developments may connect Cleveland and Detroit, Chicago and Milwaukee, all of the cities around the bay in northern California, and even the historically antipathetic cities of Dallas and Fort Worth in Texas! It is estimated than an additional 70 million people will be added to these centers of industrial expansion, partly because industry itself will be moving away from the cities into the surrounding country, and along the new belts of superhighways that make these suburban sites available to suppliers, customers, and employees.

Even automated factories will need workers, and although a larger share of the total work force will be employed in services and by the government, perhaps as many as 20 percent, or 24 million, will still find jobs as machine operators, maintenance engineers, and supervisors. All of these will be educated, alert, technically skilled, and relatively sophisticated men and women. How they will change the look of the United States is a matter of cultural and political significance as well as an economic concern. *How* will depend to a large degree on their economic situation.

Their economic situation in turn must depend upon who will be working for how long! If only 90 percent are working, and if they are working only half of the time, it is obvious that the economic situation of the population will be less than half as comfortable as it might be. Although the trend toward increased productivity per man hour may continue, and living standards may be maintained at their present levels with less work in the future, it is evident that more work at the increased rates of productivity could produce better living standards than ever before.

It has been noted already that the trend toward increased productivity has been balanced in part during the last thirty-five years by the trend toward fewer hours of work per man. If this trend continued for another thirty-five years it would result in a further reduction of the average work year by about 13.5 percent.

From 1929 to 1964, the number of hours worked per week was reduced by twelve percent, from 45 to 39.6, and the number of weeks worked was reduced by one, or by 2 percent. A reasonable guess is that the hours worked per year decreased from 2,250 to 1,940.4.

A further reduction of 12 percent in the hours worked per week would bring the work week down to slightly under thirty-five hours, and an additional week of vacation and holidays would bring the work year down to 1,680 hours.

(Hours worked in 1929 are estimated to be forty-five per week, based on figures available of 44.2 for manufacturing, and knowledge of longer hours in many other areas.)

It is interesting to note that the trend in the hours of department heads and managers is actually in the other direction, or up from 45 hours per week. Neither managers nor workers are very pleased with these trends. Although the workers are enjoying their share of the advantages of increased productivity in their pay checks, many of them would prefer to have 13.5 percent more income per year than to have 13.5 percent more free time to spend what they earn. Management on its side would rather have everyone working more regularly and more profitably for himself and for industry.

Unfortunately, industry has not as yet succeeded in finding jobs promptly for the men who could be released from their assignments as a result of new processes and automation. The record shows that except in time of war it has not been possible to put new productive enterprises in motion fast enough to keep up with the available labor supply. The result has been unemployment and a squeezing down of the hours worked per man. There is even some indication that industrial management has believed that a moderate level of unemployment was desirable as an aid to realistic negotiations with organized labor. This may have been true, but it is also true that industry cannot build new plants without knowing that labor will be available to run them. It has been difficult, if not impossible, for industry to expand in anticipation of labor becoming available! If industry must expand in an area where labor is in short supply, it often finds that it must pay a price for labor that cannot be justified by the rate of productivity. Consequently, it may be true that industry sometimes moves more slowly than it would if it were assured of an adequate supply of labor at competitive prices.

In addition to a backlog of orders, industry needs accurate information concerning the labor that can be made available, and some assurance that the cost of that labor will be in line with its productivity.

Perhaps some progress will be made in tailoring industrial expansion and the development of new markets to the quantity and productivity of the available labor in the years ahead. The problem is recognized by both industry and the government, but there has been no active demonstration as yet of any serious intent to take the difficult and radical steps that may be required to solve it.

The only trend that is apparent is that toward controlling numerical unemployment at a level equal to about 5 percent of the total labor force by permitting a reduction in the hours worked sufficient to offset the difference between the number of persons employed in new enterprises and those displaced as a result of increased productivity. If this trend continues there will be, as there are now, nearly five million unemployed year in and year out, a permanent 5 percent waste of potentially productive labor. Some economists believe that this is a hopeful trend, for it means that the total number of people employed is moving steadily upward, and that the economy is absorbing about a million new workers each year. It means that the economy is finding room for the laborers displaced by improved productivity, and is increasing production fast enough to keep up with the increased supply of labor. Moreover, it is often pointed out that the unemployed include many who aren't anxious to work, others only temporarily out of work, and millions who are relatively incapable and unskilled.

Others argue that any unemployment is undesirable inasmuch as it represents a cost to the economy to provide even a minimum standard of living for those who aren't working.

The myth that automation must inevitably create unemployment was ably exploded by Dr. William H. Peterson.[1] He points out that "the work to be done is without limit. Man's needs are insatiable," that the choice of labor is "greater leisure or more goods," and that "a reduction of hours (worked) involves a diminution in production and a lowering of real income." Although technology eliminates jobs and reduces the amount of labor required to produce specific products, it makes possible at the same time the creation of new jobs in new undertakings that become feasible and practical as a result of technology's achievements. Nevertheless, the problem of unemployment remains unsolved, and the question of whether any unemployment is desirable in practice remains unanswered.

If it were true that the level of unemployment was being controlled at approximately 5 percent of the total labor force, such a trend could be considered encouraging. Unfortunately, it is not true in fact, or at best is only partly true. The percentage of unemployed, in numbers, is being kept at the 5 percent level, and the additional workers available each year are being absorbed by the economy, but in fact the larger number of workers is offset in part by the smaller number of hours worked by all of those who are employed. Nevertheless, the trend is not altogether discouraging as long as men are

[1] William H. Peterson, *Barron's National Business & Financial Weekly*, March 5, 1962.

put to work, and per capita GNP does continue to rise.

In 1999 there will be no jobs for 6 million workers if this trend continues. It is estimated that the civilian labor force in 1999 will consist of 120 million workers, or 40 percent of the population. The civilian labor force has varied from 39 percent to 42 percent of the population in recent years. It is possible that this percentage may come down as incomes rise, retirement becomes easier, and educational requirements and opportunities increase, but it may go up to some extent as the number of young men required for the armed forces is reduced.

Of this labor force of 120 million, 6 million may be unemployed, leaving 114 million at work. If these workers only work 1,680 hours, or 13.5 percent fewer hours than they would have worked in 1964, then the work that could have been done by another 15 million workers will have been lost. In total, the combination of numerical unemployment and shorter hours will have deprived the economy of the production of more than 20 million additional workers. This could be a loss of nearly one-fifth more total production.

If this trend continues, it means that an increase in the civilian labor force of 60 percent from 1964 to 1999 will result in an increase of only 41 percent in the hours worked (from 136 billion to 191.5 billion). A potential increase in living standards of an additional 19 percent will have been lost.

In 1964, full employment (2,250 hours) would raise living standards by 23 percent, and total personal income from $420 billion or $2,210 per capita to $516 billion or $2,720 per capita, in constant dollars. An improvement in living standards of 23 percent has been lost in 1964 because it has been impossible to reach this goal of full employment.

In 1999 if the labor force of 120 million could be employed 2,250 hours, total personal income would be increased by 40 percent from the level of $1,500 billion that has been projected, assuming that the anticipated increases in productivity could still be achieved.

Personal income of $7.83 per hour worked is projected in 1999. If each worker worked only as many hours as in 1964, or 1940 hours per year, and if 5 percent of the workers remained unemployed, a total of 221 billion hours would be worked. At $7.83 per hour, this additional 29.6 billion hours would

TABLE XI

COST OF REAL UNEMPLOYMENT IN THE U.S.A.

	1929	1935	1964	1999
Population (in millions)	122	127	190	300
Workers (in millions)				
Civilian labor force	49	52.8	75	120
Employed	47.6	42.2	70.5	114
Percent unemployed	3	20	6	5
Hours Worked				
Annually, per worker employed	2,250	1,830	1,940.4	1,680
Total (in billions)	107	77	136	191.5
Potential—if all available work force worked 2,250 hours (in billions)		118.8	168.8	270
Hours lost due to real unemployment		41.8	32.8	78.5
PERCENT REAL UNEMPLOMENT	3.2	35	19	29
Cost of Unemployment				
GNP—1954 dollars (in billions)	$181.8	$152.9	$510	$1,700
GNP per hour worked, 1954 dollars	$1.70	$1.98	$3.75	$8.87
GNP LOST DUE TO REAL UNEMPLOYMENT (IN BILLIONS)		$82.764	$123	$696.295

increase total personal income by more than $230 billion, or by 15.5 percent, almost enough to double the living standards of the lowest two-fifths of the population whose income is now projected to a total of $270 billion.

As Table XI shows, the cost of real unemployment is $123 billion GNP in 1964, and may be nearly $700 billion in 1999 if these trends continue. An increase in production and in living standards of more than 40 percent could be obtained with full employment.

To summarize, the trend in the U.S.A. is toward a shorter work week, and a level of numerical unemployment hovering around the 5 percent mark. Compared with an optimum work year of 2,250 hours, there is in fact a trend toward an increasing number of unused hours, from about 19 percent in 1964 to nearly 29 percent in 1999. Whether or not this trend is desirable depends upon the relative value to the individual and to the State of more production and more leisure. If standards of living are high enough already in this affluent society, then it may be best to convert greater productivity into more time for recreation and reading and the arts. If more goods are needed to reach national goals of security, better standards of living for the lower two-fifths of the population, and improved educational and other public services, then it might be wiser to find a way to keep more of the available labor force employed at somewhat longer hours.

It should be noted, though it is evident, that a forty-hour week is not a full week of work! Most managers work more than forty hours and most workers elsewhere in the world expect to and want to work more than forty hours. This is the optimum work week of a prosperous and dynamic economy. It is a work load which could easily be raised in an emergency by as much as 25 percent!

The trend is toward an unused supply of skills and talents and energies, large enough to produce substantially more goods and services than will actually be produced. And yet, there is no indication of any slackening in the demand for more goods. It would not be difficult to find a market for the additional gross national product that could be produced with optimum employment even if all of the additional production consisted of goods for consumers. Nor is there any indication that anyone wants unemployment! Workers don't want to work a shorter week. They are glad enough to work extra hours for overtime rates of pay. Their pressure for a shorter week is a product of threatened unemployment—a device for spreading the total work over as many workers as possible. Nor does the manager want unemployment. The entrepreneur wants to produce all he can, to expand and enlarge his enterprise. The government doesn't want unemployment! Apparently, this is a trend that is moving in a direction exactly opposite to what anyone wants, but it might be a mistake to assume that no one knows how to do anything about it. It may be that the alternatives are even less attractive to some segments of the economy.

One alternative would be to create an effective demand for more workers than there are available by building a backlog of plants for new enterprises that can be put in motion as fast as workers can be found. There are always ideas at hand for new products and additional production that will be needed; what is lacking is the incentive to build the new productive facilities before they are actually required.

The problem is not just to put new enterprises in motion fast enough to make efficient use of the labor that would otherwise be unemployed. The problem is to put them in motion fast enough to take labor away from the faltering and inefficient marginal activities before it has a chance to become unemployed. The problem is to provide a fund of new enterprises which would seek out the labor that is presently used in areas of the economy that are already inefficient and are or soon will be unprofitable.

The alternative is to have idle buildings and machines that

aren't used because there is no labor available, or to have idle labor that has no work to do because buildings and machines aren't available to be used. In either case, some possible production is lost. Any amount of unemployment subsidizes some activities that are doomed to eventual failure. A similar percentage of unused resources would subsidize the employment of some labor of marginal efficiency. There is waste in either alternative, but the economic as well as the human advantages would seem to lie on the side of maintaining full employment even at the possible expense of having some extra tools and machines on hand that aren't used.

It seems likely that the only way to create these enterprises that aren't yet needed would be to put the government into business. The cost would be a competitive expense that no private business could afford, whereas the cost of unused labor is shared by the entire economy. The business enterprise pays a part of the cost in increased taxes, but it pays only a proportionate share that is not a competitive cost disadvantage. If each enterprise had to pay for the labor it wasn't using, then it would be in the same situation in regard to labor that it is in regard to unused tools and machines. So it is that the government is the first possibility as a noncompetitive source of new enterprises to be put into operation as required to assure full employment.

Another less practical possibility, but one that would be more acceptable to business, might be the initiation of enterprises by industrial associations for use as required to employ idle workers in their industries. Other more visionary ideas have already been considered. Labor organizations, fraternal organizations, schools, local governments, all might maintain stand-by businesses to be activated whenever needed to provide employment opportunities for all available workers.

To employ millions of available workers, capital will be needed for enterprises in which they could be employed. But it is feared that surplus dollars not already in use won't be available to supply the required capital. Although the labor will be at hand to build the new buildings and machines and other capital goods needed, and to operate the new enterprises, some economists argue that capital will not be found to invest in those enterprises. The conclusion is reached that some new enterprises will never be started because money will not be left over in the form of profits and savings sufficient to buy all of the new stocks and bonds that will be offered for sale. It is assumed that some labor will not be used and will remain idle because there will be a shortage of capital funds.

Although the trend in the growth rate of surplus capital seems to bear out this prediction, it will be worthwhile to consider all of the factors leading to this conclusion. One fact to be remembered is that if labor is idle for a day, it is lost forever. Whatever work was not done last week can never be done. The hours are gone, and cannot be recovered. The work that would have been done by 5 million unemployed is equal to nearly $130 million per day that is lost and gone forever. If GNP per hour worked is $3.75, in 100 days this would be $13 billion. If 5 million are unemployed continuously, this is an annual cost to the U.S. economy of more than $36 billion (at 1940 hours per year per worker). This labor would not be lost if it were used to produce the capital goods needed for new enterprises that, in turn, could employ the surplus labor made available by further automation.

What is often overlooked is that capital is not in short supply as long as labor is available. Labor is equal to capital in the end. This is often forgotten by people who think of capital in terms of money when in truth money is simply a medium of exchange, as in their way stocks and bonds are.

Machines and buildings can be purchased with money, but they can be created only with labor. The work done by labor produces a *value* and this value can be a capital value as well as a value for immediate consumption. Additional capital can be created only through work.

So it is that idle labor is the equivalent of that much capital that could have been created, and that has been lost. If idle labor were paid for not working it is evident that enough dollars would be added to the purchasing power of the economy to justify the new enterprises that could put the labor to work. Five million unemployed are, therefore, equal to a bank account equal to the value of the work that could be done by that number of workers.

Idle labor is like idle capital that isn't being used. If it is possible to go to a bank to borrow money that isn't being used, it should be equally possible to go to a bank to borrow the labor that isn't being used. Unused dollars represent an increment of value created in the goods produced by the labor that is employed, and unused labor is an additional increment of value created by the mechanized efficiency of the industrial economy. Both increments are available to be put into new productive enterprises.

A way must be found to put idle labor as well as idle dollars to work. This does not mean that a new device such as a special bank note called a *Labor Note* will be needed, but every possibility should be considered. It is almost certain that a way will be found within the economic philosophy of the capitalistic U.S.A.

In any case, a projection of the present trends points in the direction of better living standards with less work. The prospects are that in 1999 most men will give less time to their jobs than to their hobbies. They will have leisure in abundance for arts and crafts or study or music appreciation or miniature golf.

Little has been done to halt the trend toward increased unemployment. No one is happy about this trend, but business, unions, and government all seem to find it easier to accept than the possible consequences of any effective action. Waste is easy to tolerate in an economy that is still growing as rapidly as that of the U.S.A. It seems unlikely that much will be done until the U.S.S.R. or the Common Market challenges the U.S.A. in productive efficiency.

The rate of increase in the Soviet population cannot be predicted simply by projecting current trends into the future. Soviet population statistics for the last thirty-five years mean very little because the conditions that have prevailed in the U.S.S.R. during most of that time have been unique. A continuing social revolution, civil war, drought, famine, and invasion from the West have all had a depressing effect on the rate of increase. If these and similar depressants can be avoided during the next thirty-five years, the population of the U.S.S.R. may increase at an entirely different and probably faster rate.

Many factors will affect and change the current trends. If there is better housing and a higher standard of living the number of children per family may increase. If there is better food and improved medical care people may live longer. If consumer goods remain in short supply and housing continues to be cramped, if workers have more hours of leisure and have better opportunities to travel, if husbands and wives are assigned more or less often to work in different parts of the country, and if other changes occur in the living and working conditions of the people, the rate of population growth will be affected and will change in one direction or another in response to each of these stimuli.

Any measurements of population trends in the U.S.S.R. will have to be revised regularly and frequently until they begin to follow a uniform pattern that reflects a stabilized Soviet society. The records that are available do indicate that the growth rate will probably increase, and will almost surely fall no lower than its present level.

Accordingly, it is estimated that there will be 350 million people in the U.S.S.R. in 1999, or slightly more than a mathematical projection of the trend since 1929. This rate of growth is slightly less than that in the U.S.A. where an increase from 190 million to 300 million is expected. The combined population of the U.S.A. and the U.S.S.R. is expected to increase from about 420 million to more than 650 million, or by almost 35 percent. However, the U.S.A. and the U.S.S.R. together will probably account for a smaller percentage of the total world population than they do in 1964.

Whatever her increase in population, the chances are that the U.S.S.R. will be short of workers, people between the ages of 16 and 65, for at least another fifteen years. Births in recent years have not been adequate to maintain the work force in these age brackets, and it is obvious that even a command economy cannot adjust these figures at will. Consequently, the U.S.S.R. will be obliged to continue to make maximum use of all of its workers if it is to achieve its economic goals. A trend toward the employment of more young people at least part time is reported in the U.S.S.R. and it is unlikely that any significant reduction in the number of hours worked will be possible. Increased productivity per worker will be required just to reach the targets set for Soviet industry, and much greater improvements in productivity would be needed to reduce the number of hours worked.

Out of a total population of 228 million and a population over sixteen years of age of 151 million, there are 102 million workers employed in the U.S.S.R. About 45 percent of the population are in the work force and all are employed. In the U.S.S.R. it is estimated that the average worker works 2,000 hours per year. This means that about 204 billion hours will be worked in 1964 compared with 136 billion in the U.S.A. Soviet Russia is thus using about 150 percent as many hours to produce about half as many goods.

By 1999, 157.5 million workers will be working 315 billion

hours in the U.S.S.R., or nearly two-thirds more hours than will be worked in the U.S.A. The number of agricultural workers will remain about the same as in 1964, or down from 42 percent of the total work force in 1955 to 25 percent in 1999, even though agricultural production will continue to increase.

During the entire period from 1964 to 1999, nearly twice as many hours will be worked in Soviet Russia as in the U.S.A. Yet, the U.S.A. will produce about twice the value of goods and services. Soviet GNP per hour worked will have increased from $1.22 to $3.17. This is an increase of 260 percent, somewhat better than the increase expected in the U.S.A. but still not enough to bring the U.S.S.R. much closer to the U.S.A. in productivity.

Numbers of workers and hours of labor alone cannot guarantee production. In Soviet Russia the workers and the hours of work are available. What will be produced will depend on the increases in productivity that can be achieved.

These projections don't agree with the program announced to the 22nd Congress of the Communist Party by Khrushchev in 1961. The U.S.S.R. hopes to increase productivity at a faster pace than the current trends would indicate, and it plans, at the same time, to reduce the hours worked. Perhaps it will succeed. The forecasts made here are not predictions, but only reasonable projections of the current rates of growth.

15

Wages, Prices, and Inflation

IN the U.S.A. wages and prices will continue to rise as the spiral of inflation helps to encourage the annual rate of growth that is required for a GNP of $1,700 billion in 1999. This figure in 1954 dollars may turn out to be $4.4 trillion in actual 1999 dollars.

In the U.S.S.R. wage and price levels may be fixed arbitrarily but it is likely that they will be permitted to rise in response to international inflationary pressures. In any case, wage levels in relation to prices will be fixed to give the people only about two-thirds of the extra production, and to take one-third for additional investment and other governmental needs. Although total production will increase by 400 percent, a limitation on actual wage increases will limit the increase in per capita household consumption to 325 percent. However, this would be a great improvement over the previous thirty-five years, when per capita consumption increased less than 200 percent, while the GNP increased four times.

The trends are clearly in the direction of further inflation or its equivalent in both economies and, despite the qualms of many business leaders and economists, it is doubtful that any significant changes in these trends can be accomplished. It is not even certain that any changes would be desirable.

The objectives of wage and price policies should be to stimulate the expansion and growth of the economy, to accomplish efficient, profitable distribution of the goods produced, and to make full use of the available work force. If these objectives cannot be achieved with deflation, then inflation will have its way.

In the U.S.A. per capita incomes in comparable dollars are expected to double in the next thirty-five years, and in the U.S.S.R. they should increase 3.25 times. However, they may increase many more times in actual dollars if the inflationary trends since 1929 continue.

In actual inflated dollars the average per capita income in the U.S.A. may increase by another 450 percent (the increase from 1929 to 1964 was 450 percent); from about $2,630 in 1964 to $11,800 in 1999, although it may only double in purchasing power. The average income per person employed would be up to $30,000 from its present level of $7,100. The worker who makes $100 per week in 1964 will get $500 in 1999, or about $25,000 per year. The department head now earning $20,000 will get $100,000, and the man who mows the lawn and rakes the leaves will get $50 per day.

Of course, prices will go up, too. An inexpensive car will cost at least $10,000, and a pair of shoes that cost $25 in 1964, and only $5 in 1929, will sell for more than $100.

Although total personal income may be up 700 percent in current dollars to more than $3.5 trillion, prices will have increased by about 220 percent so that per capita real incomes are only a little more than twice their 1964 levels.

This is the inflationary merry-go-round that is the subject

of much discussion and controversial opinion in 1964. Almost everyone is self-consciously afraid of inflation. There is something that seems wrong with continuously rising prices just as there seems to be something wrong in raising the wage rates every year for the same job. The trouble is that there don't seem to be any right alternatives.

The theoretical alternatives are:

1. To maintain current price levels and permit wages to rise as much as is justified by increased productivity.
2. To maintain current wage rates, and reduce prices to balance increases in productivity.
3. To maintain prices and wages at current levels, and invest the profits of increased productivity in more rapid industrial expansion.

All of these alternatives suffer from the same basic defect. They can't be accomplished without arbitrary fixing of wages or prices, or both. They can't be accomplished in an economy that must respond to the pressures for higher wages, higher bonuses, more generous stock purchase plans, and higher profits, and these pressures are not noticeably abating in the U.S.A.

The first alternative is most appealing, and it is the one that seems to have the support of many leaders in business and government. Most business economists would agree to tying wage increases to increases in productivity, but few would be happy with any formula that might be devised for fixing prices. On the other side, organized labor would be reluctant to give up its right to bargain in a free market, although it would be willing to use the relationship of wages to productivity as a bargaining tool. Nevertheless, some progress may be made in this direction over the next thirty-five years.

As for the second alternative, perhaps it would be better to have lower prices and a fixed level for wages. Lower prices should open more markets and permit greater consumption, and should stimulate increased production. Theoretically, increased productivity could be reflected best in proportionately lower prices for the goods produced. In practice, this alternative would have the effect of limiting possible increases in profits, and of holding down the prices of stocks. A stable or declining level of prices would not make the owners of industry happy, for it would have a tendency to control the chances of speculative gains. Nor would it please the workman who wants a higher wage to purchase whatever goods he chooses, and who would not be as happy with general decreases in prices that he couldn't measure or see in his pay check. Nor would he be inclined to accept a fixed level of wage rates, even though he retained a right to bargain for lower prices.

The same 100 percent gain in real income would result if prices were reduced by one-half while wages remained unchanged, but it is easy to guess which direction would be preferred anywhere in the U.S.A. Indeed, who could prefer a level rate of pay, even though prices were declining at a rate of 1.5 percent per year?

The third idea is what would please management and the stockholders the most. There would be no more worries about prices or labor. Industry would just have to operate efficiently, and invest 90 percent of its profits in new enterprises. This would be a dream solution except that someone would have to fix and enforce the prices and the wage rates, and that someone would turn out to be the national government.

Other combinations of these alternative directions could be considered, but the fact is that the trend toward additional price inflation is not dictated by objective, theoretical considerations. It is a fact of the marketplace that cannot be changed by arguments or discussions. Most economists seem to believe that it will be impossible for prices to increase by as much in the next thirty-five years as they have increased since 1929. It will be impossible, they say, because it would unbalance the economy, price the U.S.A. out of world competition, and add to the precarious inflation that threatens to explode any minute. Perhaps they are right. There is, however, little evidence that the need for higher prices is being eliminated or that many powerful forces in business and labor will permit prices to be stabilized.

Whether wages and prices are higher or lower is not what matters. It is the relationship of wages to prices that counts. It is the relative rate at which they increase that determines who will have what is produced, and what and how much will be produced.

So it is not inflation of itself that matters. Price inflation has a long history, back at least to the early Egyptians, and it is probably related to the psychology of value and trade. Both sides must win in an ideal trade. Each must receive something he values more highly than what he gave up. There is thus a continuous pressure, psychological pressure, for higher prices, and the cycle of higher profits, higher wages, and higher costs continues. Both unemployment and rising stock prices are inflationary. Unless better progress is made toward deflating the stock market and putting to work a higher percentage of the available labor force than is apparent in 1964, it is likely that the trends toward higher prices will continue.

The danger that many products of the industry of the U.S.A. will be priced above foreign competition is real enough. Many *are* already. Yet the needs of American families continue to expand, and if those families receive inflated wages, they will use those wages to purchase at inflated prices all the goods they want that can't be produced in sufficient quantities elsewhere in the world. Some American prices may hold an umbrella over the prices on competitive imports. If so, the result may well be to stimulate price increases and inflationary pressures elsewhere. When labor is fully employed and maximum production is achieved anywhere in the world,

the pressures for higher wages and higher prices will inevitably
follow.

The pressures for higher prices may still prevail, but there
are some new counterpressures at work that have slowed this
trend during the last decade. The European Common Market has dramatized the factor of foreign competition, the high
cost of foreign aid points up the need for increased exports
(to get gold), and there is increasing price competition
within the domestic market.

In a free competitive economy, it was believed that prices
were *the invisible hand that guided the economy.* How will
the economy be guided now that it is necessary to weigh the
factors of productivity, wage contracts, foreign aid, income
taxes, and inflationary stock trends, as well as prices?

Increased productivity without reduced prices is inflationary of itself, and to the extent that it is not reflected in increased wages for labor, it will be reflected either in increased
dividends (the wages of capital), or, less obviously and more
discreetly, in higher prices for stocks. Increased productivity
may provide for new machines and tools, and new buildings,
but in so doing, it automatically increases the value of the
shares of stock representing ownership of that new capital, at
the expense of the public who continue to pay the same prices
for goods that can now be produced at lower costs, and the
workers who now receive the same wages for somewhat more
production. Prices could rise without inflation only if wages
and dividends increased in an amount exactly equal to the
increased revenues created by the higher prices.

If price inflation is coming to an end, then stock market
inflation and wage inflation must come to an end at the same
time. The end of inflation would mean the end of the speculative boom in so-called growth stocks. The factor of appreciation due to inflation would be eliminated and most of the
growth in growth stocks would be eliminated with it. The
stock market, minus inflation, would be calm and collected

and not interesting or profitable for speculators. Stock prices would not rise automatically over a period of a few years, earnings would not multiply, and profits would rise and fall within a short range, dependent on the skills of management in developing production and marketing efficiencies.

In an economy without the overly optimistic factor of inflation, the fluctuations in earnings from industry to industry, and from business to business, must be measured against a base of zero, whereas in an economy that is enjoying a continuous rate of inflation, all progress is measured against an automatic gain of 4 or 5 percent. In an inflationary situation even a badly managed company can report a modest profit, whereas without inflation the same company might be on the verge of failure.

Inflation does stimulate investment because of the extra growth incentive it offers to every investor, the hedge of 4 or 5 percent against any possible loss. It may even be a desirable influence to persuade potential investors to forgo immediate consumption in favor of investments in new enterprises. What inflation does is to take from the creditor a percentage of his loan and permit the debtor to reduce his debt by the same percentage. Thus, continuous inflation inflicts a penalty on all those who have loaned their resources or their labor at fixed rates not adjusted to inflated values. And it offers an advantage to those who can borrow either labor or resources on the same terms.

If inflation continued at its present rate for thirty-five more years, the net effect would be to pay excess profits continuously out of rising prices that remained always a step ahead of rising wages and costs of materials. Prices would double or triple again, and the increased productivity of the economy would be divided more or less unequally among the owners, managers, and workers with a larger share going to the borrowers, mostly the owners, and a smaller share to the lenders, including the managers and workers.

If inflation can be controlled, and prices can be adjusted to increased productivity, the effect will be to maintain wage rates and reduce prices, or to maintain price levels and to raise wages in proportion to increases in productivity. The worker, the manager, and the stockholder will, in effect, be assured a fair wage for their work and their capital investment, but the speculation will be taken out of unreasonable demands for increased wages, managerial bonuses, and profits.

What would happen to the incentive to build a business if profits were uniform, values stabilized, and stock prices firm?

The man with imagination, ambition, and a good record would be able to get capital for his business, but he would have to pay a modest rate of interest, and would be able to earn only a percent or two more than he paid. His income as a manager would be proportionate to the size and complexity of his enterprise, and he would live well enough, but it is unthinkable that he would ever be able to return the capital that he borrowed. The small amount that he could earn in excess of what he would have to pay for the capital invested in the enterprise would be the equivalent of a modest incentive payment for his good management.

If speculative gains became nearly impossible in an economy with stabilized profits and prices, and insured risks, daring and resourceful men would seek out the large and difficult jobs for which they could hope to be well paid. Those who already had large fortunes would find their incomes reduced as profits became smaller, and would see their fortunes shrinking as taxes multiplied and opportunities for capital gains disappeared.

In the long view it is the capitalists and the risk-takers who will find it in their self-interest, financially, to resist the extension of automation, scientific procedures, and electronic controls to the economy as a whole. Once all of industry is operated scientifically, there will be no need for risk-takers because there won't be any risks. The excursions into outer space that have been controlled with mathematical precision by the N.A.S.A. are good examples of the scientific way of going that can be extended to every kind of economic activity. Individual men will always take calculated risks, but the chances will be known in advance, and with a certainty that success must follow a predetermined series of tries. When industrial management is reduced to a science, the supply of materials and workers alone will determine what new enterprises can be undertaken.

It is evident that an economy without potential inflation would not be welcomed by anyone. The investor and the entrepreneur would cry out the loudest for their gains would be cut the most. Managers who were paid for managing might suffer the least, but it is unlikely that they would willingly give up the chance of sharing in profits. Workers at fixed rates for their work would probably not be pleased even though they might receive their share of increased productivity in reduced prices.

What is more serious is that the rate of economic growth would probably be inhibited if the incentives for new enterprises and expanded production were limited to a modest rate of interest on the additional capital put to work or a share in the increased productivity exactly equal to the contribution made by the capital invested.

In the U.S.A. it is likely that continued growth will require increases in prices and wages together at a rate that will permit the investment of increased profits in expanded production.

The chances are that productivity will continue to increase somewhat faster than incomes, as it has during the last thirty-five years, and that increased debt supported by increased federal spending will make up the difference between incomes and prices.

The general tax reduction, passed early in 1964, has the

blessing of industry and is another more devious method for achieving inflation without increasing prices and wages coincidentally. If it is followed by a reduction in prices, further inflation can be avoided, but it is not intended for that objective. It is expected to give corporations more funds for reinvestment in new capital and, since it extends to the lower personal tax brackets, to give consumers with unsatisfied needs more purchasing power. To the extent that it is weighted in favor of corporate taxes, it amounts to a support for sagging markets in corporate stocks without price increases. It is thus a device for blowing up one side of the balloon faster than the other side, but it will not obviate the pressures for price increases and wage increases for very long.

It is almost certain that some inflation will continue through the next thirty-five years, though prices may increase only 150 percent instead of 300 percent, and wages may increase only three times instead of five times.

It is also apparent that some inflation may be desirable and even necessary in a capitalist economy to provide incentives to industry to increase production at a rate fast enough to keep up with the supply of labor, and to produce the goods needed to maintain and extend a rising standard of living. Without inflation it is doubtful that industry in the U.S.A. will increase production fast enough to keep up with the opportunities offered by automation, new technology, and the total labor force available. Even with inflation, it is obvious that some additional production could have been achieved in the last few years and it is likely that some form of inflation will be around for many years to come.

In the U.S.S.R., where prices are controlled arbitrarily, wages can be permitted to increase in proportion to increasing productivity. If per capita household consumption increases by 325 percent in the next thirty-five years, then per capita wages might increase by the same amount. A school teacher who earns $100 per month now might get $325 per month in 1999.

Prices are fixed in Soviet Russia, and although they can be adjusted as required, they are firm enough to permit the inclusion of the printed price on the labels and packages that are used for candies, cookies, beer, and many other items. In Krimsk, I learned that the prices to be paid for supplies and raw materials are also fixed and remain unchanged for several years at a time. The farmers who raise corn and beets and cucumbers know what prices they will be paid for the goods they will produce over the next two or three years. As the director of the canning plant in Krimsk said, "This makes it easier for everyone to make his plans for the future." Moreover, prices on finished goods are the same from one town to the next. On canned foods and beer I was told that there are only three or four price zones in the country, and that prices in the same zone are identical.

Wage rates are set in much the same way, and as total production for household consumption increases, wages can be increased everywhere in the same amounts. There is no need for collective bargaining in the U.S.S.R. for rates are pretty well stabilized by industries and jobs, and adjustments are made almost automatically to keep up with any increases in the production of consumer goods. Industrywide national bargaining is the last thing that a labor leader should want. In such a situation he would have to accommodate himself to the meaning of all the relevant economic facts.

In the U.S.S.R. management can take whatever rate of profit it needs for investment in new industrial units. Since 1929, its profit rate has been high by American standards, but it has come down somewhat and it may reach a level of only 20 percent in 1999. (American capitalists could argue that if they were given the rate of profit enjoyed by industry in the U.S.S.R., they could easily expand the economy at an annual rate faster than that of Soviet Russia. However, they might have an insurmountable marketing problem.)

This assumes that the percentage set aside in Soviet Russia for investment is comparable to the profit in American industry that is available for reinvestment. Although this isn't exactly so, it is true that Soviet prices and wages are set to produce a profit of 25 percent or more that can be used for new capital expenditures.

If prices in the U.S.S.R. are permitted to increase over the next thirty-five years, then wage scales will have to increase even faster to keep up with the increased supply of consumer goods. If prices should double, for example, wages would have to increase by 650 percent, and the schoolteacher with $100 a month in 1964 would have to get $650 per month in 1999.

Although an economy like that of the U.S.S.R. can control prices absolutely, it turns out in practice that it is often better to let prices follow world trends, and to control wages instead. Otherwise, there would soon be two sets of prices, one for domestic use and another for international trade.

If prices were stabilized in the U.S.S.R. while prices elsewhere increased, the effect would be to inhibit foreign trade by making rubles too expensive in terms of dollars or other foreign currencies. The tendency has, therefore, been to let Soviet prices follow the inflationary trends of world prices.

Rising wages reflect the increasing productivity of Soviet industry and the allocation of larger amounts of total production to consumer goods. Wage increases are planned to equal exactly the value of the increases in the production of goods for household consumption. Planned increases in production precede or accompany planned increases in buying power. Wages and prices are designed to make sure that all goods available are consumed.

Another function of wages in the U.S.S.R. is to reward and provide incentives to workers in proportion to their contributions to the total objectives of the State, not just to the economy. They are set to distribute all the goods available for household consumption to selected segments of the population with more regard to ability than to need.

Because there is no open market for labor, wages cannot reflect with any accuracy the relative value to the economy of specific jobs and skills. At the present time, the spread in wage rates within industries is wider than in the United States. If this spread continues, many workers will receive considerably less than the average in 1999, and others will be paid much more. The actual range of incomes will probably extend from a low of $500 to a high of more than $5,000, assuming that the average is $1,785—a range still narrower than that which will exist in the United States.

This wide range in wage rates within an industry strongly suggests that the pressure for maximum production has tempted management to offer greater incentives than could be justified economically to selected groups of workers whose skills were essential, and who could be persuaded to take the lead in setting a fast pace for the industry. The consequence is that workers at the lowest levels are sometimes underpaid while those at the top are overpaid. The tendency is thus toward a less equitable distribution of goods than in a market economy where wages are set more rationally. In this case it would seem that the tools of communism don't lead in the direction of communism. They even lead away from communism.

If it should turn out in the end that the workman is worthy of his hire, and that those who are overpaid will be discovered while those who are underpaid will get their due, the trend will still be away from the precept of communism that each worker should contribute what he is able and receive what he needs. It will be in the direction of giving each worker a share of the total in proportion to his contribution, whereas at present he is getting back either more or less than he contributes.

According to Communist theory, the least productive workers should be the ones to receive shares of total production larger than their contributions, and the best workers those who would be satisfied with shares smaller than the value of their contributions. In practice, the opposite seems to be the case in the Soviet Union as in the U.S.A. The better worker who is more efficient and more industrious is valued beyond his actual contribution. He is essential to the success of any attempt to expand the economy, and it is recognized that he has the option of working as hard as he can, or of taking it just a little easier. To get his maximum effort an appeal must be made to his ambition, his pride, and his desire to have a better standard of living for his family. Consequently, he is offered inducements, incentives, and extra rewards.

The less efficient worker who is lazy and lacking in ambition is valued at less than his actual contribution. He is a symbol of the past, of the drag on the Soviet economy from the uneducated bulk of the peasant population. In any case, he can be forced to do his share of the work. He can be driven extra hours at tedious jobs and paid with a minimum ration of bread and cheese. He is the draft horse who can be worked and worked and worked as long as he is given his regular bag of oats. Although his work is essential in the aggregate, his individual contribution is small, and he can be paid even less than he is worth.

During the next thirty-five years it is likely that the present relatively wide range in wages will be narrowed somewhat as training and education are provided for all workers and as upper income levels become stabilized at comfortable to affluent standards of living. It is also likely that minimum living standards will be raised substantially through additional communal benefits. Minimum requirements of food, clothing, housing, medical care, and education may well be supplied by the State without charge. Money wages may still vary widely, but there will probably be a somewhat more equal distribution of the total production of consumer goods.

Although Soviet prices and wages are set arbitrarily without the aid of a free marketplace, the valuable effect of a market is not entirely lost. Prices are set to dispose of the goods available for sale. Because prices are not necessarily related to costs, and because costs cannot be determined with any accuracy as long as wages are fixed outside the marketplace, they may be set too low so that shortages occur and the demand or need for such items is magnified, or they may be set too high so that other items seem to be unwanted and overproduced. Consumers do have some money to spend and they can choose whether or not to buy the goods offered at the prices asked. If some items are priced too low, these will disappear very quickly, and will be offered for resale at higher prices in an unofficial and illegal black market. If other items are priced too high they will not sell, and their prices will have to be reduced until they can be sold.

Consequently, the effect of a free market is obtained slowly and with some difficulty as far as prices are concerned as long as a black market is permitted. However, unless there is a black market for labor, the demand side of the marketing equation is still lacking, and the marketing situation is still artificial and unrealistic.

It is hard to tell how much of a black market for labor exists at present. There are some indications that labor may move from one plant or one industry to another in response to offers of better wages or better living conditions, and stories of moonlighting persist. It is almost certain that such a black market will grow as workers become better trained and as the shortage of workers continues. Soviet economists may thus get the benefits of a free market without giving up their ability to impose arbitrary controls when they feel these are required.

The inflationary trends in the U.S.A. and the U.S.S.R. are not discouraging from an economic point of view. Deflation comes automatically when consumers don't buy and prices sag, and with it comes increased unemployment and decreasing profits. Inflation is equally automatic when consumers are buying everything there is for sale, and the problem is to produce additional goods fast enough.

Some sort of inflation would seem to be inevitable in an expanding industrial society. It is hard to tell whether expansion brings about inflation, or whether inflation contributes to expansion, but the two do seem to go together. In an expanding economy a large amount of work must be invested in future production and the payment for that work, even if the payment is deferred as in Soviet Russia, constitutes an inflationary pressure.

In the U.S.A. increasing demand stimulates the production of additional goods, and in the U.S.S.R. additional goods are produced according to the national plan, and then wages are increased to create additional purchasing power. The difference, though vital to Communist and capitalist thinking, is not significant in its effect. In both cases additional goods are produced and consumed, and a share of every worker's yearly pay is invested in expansion, whether he likes it or not and whether it is obtained through higher prices or lower wages.

It is true that inflation hurts those with fixed incomes who don't have work to invest in future production. However, it is probably unrealistic to expect that any plan for limiting inflation can succeed. The pressures for continued industrial expansion will probably prevail.

Without some price inflation, the increased productivity that creates increased supplies of goods available for consumers is reflected in lower costs per unit produced and a larger margin between selling prices and costs. If that increased margin is divided between the workers, managers, and owners, it results in higher wages and salaries and a small increase in profits. However, in total it is just equal to the price of the additional goods produced, and is therefore just large enough to purchase those goods. It doesn't create any surplus to invest in expansion.

If prices are increased, there is a larger amount to distribute and the chances are good that a larger share will go into profits.

16

Debt and Government Spending

DEBT represents the hours of work that are invested in the future, the hours that are used to produce machines and tools that can't be consumed in the present and that must be paid for with promises of future rewards. It also represents the value of things that are needed for use now and that must be paid for with work that will be done in the future.

In the U.S.A., where each individual can choose whether to invest his work in the future or use it entirely for his current needs, a value is put on his willingness to wait until sometime later for his pay. The man who invests in the future can expect to be paid a small amount each year for waiting, and in the end his total pay may be more than twice what it would have been if he had been paid off immediately.

In the U.S.S.R. the individual can't choose whether or not he will invest his work in the future. A certain share of his work is commandeered. Nor is he paid at all for the work he has given up. He receives no yearly interest nor any eventual repayment. He is like the farmer who works extra long hours to improve his land with the hope that sometime in the future his crops will be larger and his work easier, but he has no guarantee that this will be so.

There is work invested in both the U.S.A. and the U.S.S.R. and in a capitalistic society, that work, being immediately nonproductive, is offset by debt. Consequently, it is not illogical to attribute to Soviet Russia a debt that would be equal to the value of the work invested in future production, though there may be no evidence of any such debt.

What matters is that debt is a useful and essential tool for financing economic expansion. Debt or its equivalent must be incurred to get the machine in the first place, and to keep improving and adding to it. In effect, debt is an investment in the economy, and it is inevitable that as the economy grows debt will continue to grow.

In the U.S.A. debt has been about 185 percent of the GNP since 1929 with only modest fluctuations except for an exceptional increase to 300 percent during the Depression. If it is 185 percent of the GNP in 1999 it will exceed the $3 trillion figure. It is estimated that government debt will be about $786 billion, more than double the 1964 ceiling, even if it continues its recent downward trend as a share of total debt. Table XII shows a comparison of the debt in 1929, 1964, and 1999 in 1954 dollars. With continued inflation, the actual figures for 1999 would be much larger.

TABLE XII

DEBT—BILLIONS OF CONSTANT DOLLARS—USA

	PUBLIC	PRIVATE	TOTAL	% OF GNP
1929	$ 29.7	$ 161.2	$ 190.9	183
1964	350.0	780.0	1,130.0	188
1999	786.25	2,358.75	3,145.0	185

Public debt was 66 percent of the total debt in 1945 at the end of the war, but it has come down to 31 percent in 1964, and it is very possible that it could drop below 25 percent in another thirty-five years. Such a further decline is unlikely, however, as long as there is a continuing need for the government to stimulate the growth rate of the economy and to prevent unemployment from rising much above 5 percent.

This is a modest prediction of the amount of debt that will be required to finance an expanding economy of $1.7 trillion in 1999.

Government spending for capital improvements, roads, and research; private investment in new productive units; and further increases in consumer credit will all be needed to maintain a growth rate of 3.5 percent annually in the GNP.

This increase in total debt amounts to an additional investment of $2 trillion that will be needed over the next thirty-five years to finance the projected increment of more than $20 trillion worth of production.

It is estimated that government expenditures in 1964 will amount to $90 billion, or 15 percent of the GNP (current dollars). If government spending were 15 percent of the GNP in 1999 it would total $255 billion (1954 dollars) but this would mean an unlikely increase in expenditures for national security from about $50 billion (or $42.5 billion in constant dollars) to more than $140 billion. Although the cost of security may continue to rise despite the increasing efforts to move in the direction of disarmament and a reduction in the costs of nuclear tests and missile production, it would be overly pessimistic to guess that no progress will be made in controlling military expenditures. It would be equally unrealistic to assume that the $42.5 billion level can be maintained or even reduced, although such an objective might be desirable. It is possible that the annual expenditures for security in another thirty-five years will not be much more than half again as large as they are now. A total of $70 billion in 1999 is at least a reasonable guess.

The direction of the trend in government expenditures is hard to predict. A conservative view may be that the federal government will continue to spend more dollars but that *if* the rate of growth of 3.5 percent is maintained, its spending as a percentage of total production will come down. This is an important *if*. If a 3.5 percent rate of economic growth is not maintained, it is fairly safe to predict that government outlays will multiply in dollars and as a share of the GNP. If there is increasing unemployment, the federal government will do something about it eventually, and whatever it does will be costly. And if the economy should lag behind the growth rate achieved in western Europe and the U.S.S.R., the government will undoubtedly do something about that, too. There are many things it could do besides subsidizing military research, including aid, investments, and projects for expanding the use of national resources, all of which would be expensive stimulants to the economy.

That such expenditures will more than double their present levels by 1999 is a fair guess. If so, a total of about $200 billion would be equal to 11.8 percent of the GNP, down from an estimated 15 percent in 1964. There is little indication that either federal spending or inflation is going to be controlled. Only the government wants to control inflation, and only business wants to control government spending.

In the days of the New Deal when large amounts were going into experiments like the W.P.A. and the C.C.C., this was called "priming the pump." Now expenditures for war materials, research in outer space, public works, and many other projects too vast or too speculative to be undertaken by private industry are accomplishing similar objectives. All of these together amount to "pump priming" on a spectacular scale that has probably helped to stabilize the economy. Without these unusual commitments amounting to more than half of the federal budget, or about 8 percent of the gross national product in recent years, it is likely that there would have been severe unemployment and a serious slowing in the annual rate of economic growth in the U.S.A.

Even with a federal budget amounting to 15 percent of the GNP, the annual rate of growth slipped to an average of only about 2 percent for the years from 1955 to 1961, and a sharp reduction in such spending could make the difference between continued growth and a dangerous economic depression. It seems obvious that federal expenditures in these proportions could continue indefinitely.

An adequate military organization amply supplied with the most modern weapons will be maintained for many years to come. Military force will be needed and will be used to overcome and contain irrational and hysterical situations in the world. Whether it is U.N. force, U.S.A. force, or U.S.S.R. force, it is almost certain that it will be employed for keeping the peace and for gaining or retaining limited competitive advantages. Expenditures in the U.S.A. for defense or in preparation for war will continue and may even increase in total dollars partly because defense is still necessary and partly because defense contracts contribute to maintaining employment and profits. Although the chances of a nuclear war may continue to decrease as it becomes more and more apparent that such a war would be wholly unprofitable for all concerned, it is still likely that many war industries will continue to flourish.

As long as the war industries don't take workers away from other more useful and profitable activities it is evident that they are making a valuable contribution to the economy. The dollars spent to produce missiles and rockets may not immediately improve the standard of living, but those dollars are respent again and again during a year to purchase the things that are desired by the people employed and to stimulate the production of all kinds of additional goods.

The question is not whether the war materials and other goods and services purchased by the government are essential, but rather whether the employment and profits resulting from government expenditures would be lost without them. It seems likely that a long period of adjustment will be required to substitute private production and employment for a significant portion of the activities now managed or subsidized by the government. What is needed is marketing vision of a high order to stimulate new production that will make efficient use of all the available resources of manpower and capital. It is even to be anticipated that the government will have a share in the risk of new marketing adventures that may extend around the world. The part played by the government will depend on the capacity and imagination of sales and marketing managers and the resourcefulness and daring of large and small businesses all over the U.S.A.

Whatever the level of government spending, a total debt in 1999 of more than $3 trillion is a realistic projection of the present trends. The interest on the public debt alone will amount to more than $28 billion, or about 14 percent of the federal budget.

In the U.S.S.R. the capital investment made each year has increased from an estimated $8 billion in 1929 to $68 billion in 1964. In the last thirty-five years a total of at least $1,000 billion has been invested in the Soviet economy.

In the next thirty-five years the annual rate of investment is projected to increase from $68 billion to as much as $200 billion, and the total additional investment to be made will probably amount to more than $4.5 trillion.

If the total debt is considered to be about equal to the total investment in the Soviet economy, then the "debt" of the U.S.S.R. would be $1,000 billion in 1964, and would reach a total of more than $5,000 billion in 1999. Although this "debt" will never be repaid, it does represent the amount of work invested by the people of the U.S.S.R. in their economy. They may not get any return, as such, on this investment, but they might argue that their return can be measured by the improvements they have gained in their living standards.

The "debt" that is owed to the people of Soviet Russia could be calculated more accurately in capitalist terms if the total value of the Soviet capital, the factories and tools and warehouses that comprise the industry of the U.S.S.R., could be determined. In theory, this should amount to the total cost of building and maintaining Soviet industry less allowances for depreciation and obsolescence. In this same sense the total debt of the U.S.A. would be equivalent to the value of all government and corporate stocks and bonds. In the U.S.A. the amounts invested in the economy are represented by these promises to pay and certificates of ownership. These investments represent work that has been done but that can't be paid for with goods available for immediate consumption. As in the U.S.S.R., a certain amount of work is needed for investment in the future, but in the U.S.A. the worker is paid for all of his work, and then is persuaded to buy enough stocks and bonds to pay for the new capital goods that were produced. "Worker" is used here to mean anyone who works or receives an income, and it is understood that the workers who can buy the stocks and bonds are those who have surplus incomes after providing for their immediate consumer needs.

Although the Soviet worker is not paid at all for the work he is required to invest in his new industrial economy, and has no stocks or bonds to show for the share he has contributed to and might be said to own of that economy, he is, nevertheless, in a position similar to that of the American worker. He is a creditor of his economy, and he has a reasonable prospect of getting back a fair return on his investment. His standard of living has already improved, and as production continues to grow he has reason to be confident that he will continue to get some share of the new products that are being produced.

However, he should not hope to get a share in proportion to the work he has done, or the contributions he has made to the success of the Soviet enterprise. As in the U.S.A., some influential and powerful segments of the population will get more than others in increased wages and fringe benefits if not in dividends and interest payments.

The profits will be paid in full in increased standards of living for all but it is unlikely that they will be paid more fairly or equitably than elsewhere in the world.

In the U.S.S.R. the workers have given up from 30 percent to 40 percent of their work each year since 1928 to build an economy that could some day give them all a decent standard of living. They will probably continue to invest 25 percent to 30 percent of their labor for another thirty-five years, but they should soon begin to reap a better return on their investment. It has already been pointed out that living standards more than doubled from 1929 to 1964, and that they are likely to increase another 400 percent by 1999.

It is sometimes argued that the total government debt of the U.S.A. is no problem because it is owed to the people of the U.S.A., and this may be true, although the individuals who owe are not identical with those who are owed. In the U.S.S.R. the total "debt" is indeed owed to all the people. Five thousand billion will be owed in 1999, an amount that represents the value of what will have been given up by all the people, and what should be repaid with interest to them. In this sense "owing" means recognizing the deprivation, the hard work, and the sustained belief that enabled men to do the things that had to be done to create a trillion-dollar economy. Many have been deprived and will be deprived during the decades ahead, all will have to continue to work hard, and all believe that they will be paid the promises of the Communist ideology. How they are paid will determine in part whether the Soviet goals will be achieved and will reveal what communism turns out to be in practice.

Mounting interest costs in the U.S.A. are paid to those families that had surplus earnings to invest in the growing economy. Thus, they continue to receive a somewhat larger share of the GNP than those whose incomes were only enough or not enough to provide for their own needs. The owners and managers of the economy are assured larger portions of its bounty than ordinary craftsmen and laborers.

In the U.S.S.R. the more generous allotments of the total production will probably go to those who have invested their talents and risked their lives to create a communist political economy. The commissars and political leaders are equivalent to the owners in the U.S.A. but they have less assurance of any continuing advantages. In the U.S.S.R. whoever gets more must prove his ability and skill from day to day whether that skill is political or economical. The measure of the Soviet economy and its dedication to communism will be found in the distribution of production that is achieved in fact in 1999.

17

Automated Production

IF the trend toward automation continues, it is more than likely that the productivity of industry in the U.S.A. will increase faster than the mathematical trends in the gross national product indicate. The technical skills are available, the machines and controls are on the drawing boards if they are not already in limited use, and there are sufficient competitive incentives to urge industry toward maximum productive efficiency. If more complete use can be made of the available labor force to produce the additional machines and goods needed to satisfy the demands of the people of the U.S.A. for a more abundant standard of living, the possibility of increasing the number of hours worked and the productivity of those hours might be realized.

A conservative projection has been made that productivity per man hour will more than double in the next thirty-five years. But it could increase by as much as 400 percent. The problems of unemployment are not being solved, and the trend toward automation will be limited by the willingness of industry to invest large capital sums in new plants and new processes. The production of steel in automated continuous production plants is already a reality in Europe, although it is still considered to be economically impractical in the U.S.A.

The progress toward automation in the U.S.S.R. may persuade the U.S.A. to move ahead more rapidly. In the U.S.S.R. there is no problem of finding work for the men displaced at all levels by computers and automatic machines. The U.S.A. cannot slow down its pace toward automation in order to gain time for solving these problems without giving the U.S.S.R. the opportunity to overtake and go ahead of the United States in production efficiency. However, if we in the United States don't slow down, our problems with labor will be enlarged and complicated. This is a quandary we cannot avoid in the next few decades.

The possibilities of automation are apparent, and in some industries examples of what it can do will be revealed within another thirty-five years or less. In some factories the entire production process will be controlled with electronic computers that measure each step and make continuous adjustments to cope with every problem and variation in the rate and quality of production. The time will come, and very possibly before 1999, when mechanical programming of production processes combined with automatic communication between the units of a world-wide enterprise will be so perfect that very few human beings in any factory will know what is going on there! Certainly the President and the Vice-President, Finance won't know. A planning headquarters situated anywhere in the world will run an entire multiplant business using formulae stored on microfilm that will be transmitted electronically, together with current economic factors, to widely separated operating divisions. Producing units will be so neatly tied together and coordinated through the programming center that a breakdown in a factory in Denmark will automatically increase production in a similar plant in Peru! The meaning of productivity in such an enterprise will be understood to be a function of the quality of the raw materials and the efficiency of the tools, buildings, processes, and labor required for each productive unit, and productivity will be measured accurately and recorded on film for every fragment of the total activity. Electric eyes and clocks and color cameras will note automatically how well and how quickly every job is done, and the advantages and improvements achieved in one area will be transmitted immediately to all other units performing the same function.

These and other possibilities of automation will be explored in the next thirty-five years. The pressure of potential automation will be felt with increasing weight in the decades ahead and, in combination with the pressure for increased employment, may force a breakthrough to new economic levels that can only be imagined now, and that will probably be reserved for the century ahead.

Automated industry may evolve more rapidly in the U.S.S.R. than anywhere else in the world. Nearly all of the conditions in Soviet Russia are favorable. Some of these are:

1. *Supply of labor.* A labor shortage is anticipated for the next twenty to thirty years. Labor will not oppose automation and almost certainly will favor it.
2. *Quality of labor.* Although the total labor supply is limited, an increasing number of young people are being trained for the jobs of operating and supervising highly mechanized production units.
3. *Available resources.* A large share of the Soviet GNP is already earmarked for investment in new capital projects. New automated factories can be built without cutting back the production of consumer goods.
4. *Cost.* Most of the Soviet industrial plants are in need

of modernization or replacement. It will be nearly as easy, only a little more costly, and much less expensive in the long run to convert them as completely as possible to automation.

5. *Power.* Electrical power is becoming available in Soviet Russia in the large quantities that will be required for automation.

6 *Machines needed.* Although the tools of automation, machines for continuous production, electronic controls, computers, and data-processors are not available in unlimited quantities to Soviet industry in 1964, they can be obtained elsewhere in the world, and they will be produced in increasing varieties and quantities in the U.S.S.R.

7. *Long-range objectives.* Automation is completely compatible with the long-range objectives of the economy, which seem to be to increase total production as rapidly as possible, to create the most efficient industrial economy in the world, and to raise living standards at the same time.

The lack of trained workers and the difficulty in getting enough of the most modern machines are the only factors likely to retard the pace toward automation.

The brewing industry is a good example of the opportunities there are in the U.S.S.R. to move toward automation at a cost that is actually less than the cost of bringing the present operations up to today's standards. To mechanize and modernize the 427 breweries in the U.S.S.R. would mean the replacing of thousands of wooden tanks and fermenters, miles and miles of piping and conveyers, the rebuilding of hundreds of refrigerated cellars and storage areas for supplies, and the addition of new machines for measuring, weighing, and controlling the many steps in the brewing process.

The alternative is to move immediately to new automated

continuous brewing operations that appear to be nearly ready for practical application. This is what the U.S.S.R. seems to have in mind. By re-equipping present breweries with elements that fit the new concepts of continuous brewing operations, these plants can move toward almost completely automated production of beer in more compact and less costly facilities. Because the Soviet brewing industry has lagged far behind the rest of the world, it is now in a position to move all at once to the most advanced brewing techniques at a cost that will be much less than the cost of just catching up.

There is every reason for Soviet industry to become mechanized as rapidly as possible. The most efficient methods of production together with maximum employment of all of the workers who can be recruited into the labor force will be needed to reach Soviet objectives. Much progress has already been made but Soviet industry in 1964 is still at a level comparable to that of the U.S.A. in 1929. Soviet factories vary greatly in regard to the numbers and the quality of the machines they are using. Many plants were built in the 1930's or before. From 1929 to the end of World War II many new factories were built, but some of these were destroyed during the war. Another substantial share of the total including all of the most productive and modern units were built in the last twenty years.

It is estimated that the total cost of the Soviet investment in industry during the last thirty-five years was only slightly less than the amount invested in new industry in the United States during the same period of time. This investment was little enough when it is measured against the task that remains of creating an entirely new Soviet industry. The amount spent in the U.S.A. was what was required to maintain and expand an industry that already existed. Less than that amount was available in the U.S.S.R. to do the whole job. It is impossible to find out exactly how much was accomplished, but if there is a modest slice of Soviet industry that is as new and efficient as any units of industry in the United States, there must be another much larger share that is obsolete and worn out.

For the present and even for another ten years this may seem to be a disadvantage, but if additions to capital investment continue to increase each year, the time may come when the Soviet industrial plant in total will be newer and more efficient than that of the U.S.A. It is estimated that the U.S.S.R. will invest another $4,020 billion during the next thirty-five years, while the U.S.A. is investing only $3,375 billion. If these estimates are even approximately correct, the U.S.S.R. will be building and rebuilding its industry at least 20 percent faster than the U.S.A.

Despite these trends and probabilities, productivity in the U.S.S.R. is not expected to exceed much more than one-third the rate of productivity in the U.S.A. This apparent paradox is explained by the fact that a large share of the workers in Soviet Russia are still employed in agriculture, from 40 percent in 1964 down to an estimated 25 percent in 1999. It is calculated that if a GNP of $1,000 billion is achieved, the GNP per hour worked will be only $3.17 compared to $8.87 in the U.S.A. If the shift from agriculture to industry is faster than is anticipated, then the hours worked may be fewer, and the productivity per hour higher; or it may be that the gross national product may far surpass the $1,000 billion figure.

It is also possible that the rate of growth may not be maintained, and that the amounts available for investment in capital goods may be cut back proportionately. The GNP may reach a level of only $800 billion. In this case it is still unlikely that the trend toward automation would slow down. The need for maximum productivity would still exist, and all of the reasons favoring mechanization would still be valid. The result would probably be a less rapid rebuilding of industry, accompanied by a slightly lower rate of productivity and some reduction in the hours worked.

It is not speculation to predict that industry in the U.S.S.R. will continue to move in the direction of complete mechanization and increased efficiency. The shape of a new kind of industrial economy will begin to be visible within the next ten years in the U.S.S.R. What that economy will mean to the people of the U.S.S.R. and to the people of other countries is a speculation, but it is an estimate that must be made with the greatest care before any realistic plans for the future can be completed anywhere in the world.

18

Marketing and Distribution

THE trends of marketing in the next thirty-five years could be the major factor in determining how large a GNP can be achieved by 1999. How many of what products will be produced will depend upon marketing's success in finding customers for the many things that might be produced. Thus marketing is the key to the high national standards of living that could be attained with full employment in automated factories.

The tools and the manpower are available to produce the goods if marketing can find out what goods could be produced and distributed at a profit. If marketing can solve the reciprocal equations that would be required to tell what goods would be needed by the persons who would be employed in producing all the goods that could be produced, it is entirely possible that total production in 1999 will be substantially more than $1.7 trillion.

Mathematical trends are of little use in projecting what

marketing may be able to do in the next thirty-five years. They point to the same total levels of sales that are indicated by the trends in production. What is produced is sold and distributed somehow, and there is no way to separate these trends. What is needed is the trend showing what could have been sold! This can be estimated by looking at the progress that marketing has made in physically distributing the goods that have been produced, and in evaluating the immediate and potential needs for additional goods.

Marketing's efficiency in retail and wholesale distribution has been described already. If this trend should continue for another thirty-five years, mechanized warehouses would be the rule, and unmanned trucks piloted by radar would be conveying goods of all kinds from New York to Denver, and from Seattle to Chicago, as smoothly and easily as they are lifted and carried around a factory today. Wholesale distribution would amount to an extension from the factory to the retailer of the automation that would be producing a continuous stream of standardized products.

Retail distribution, in turn, would doubtless have continued its progress toward electronic shopping, probably *sans* pushcarts and *sans* money! Retail credit cards are already almost everywhere, and the possibilities of shopping by sample either in air-conditioned traveling show-carts or by TV are at least being discussed and considered. Double and triple costs of transporting goods, stocking and restocking shelves, and handling, displaying, pricing, and counting the same goods over and over again offer opportunities for savings in the cost of retail distribution that could save the consumer as much as 10 percent of the total price he has to pay. (Assuming average retail markup is 20 percent, a saving of one-half in the cost of retailing would have the effect of increasing living standards by 10 percent!)

Despite the many changes that have been made since 1929, marketing has in fact lagged behind production in its progress toward maximum efficiency. Production efficiencies have reached the point where additional improvements in the production process generally contribute only fractional savings. Meanwhile marketing costs are rarely broken down on a per unit basis and in fact are rarely measured at all except in terms of the total success of the total marketing activity.

The difficulty in measuring the efficiency of separate marketing operations is apparent. Advertising and promotion, and even selling, are personal activities that can't be measured mechanically. It is impossible to say what factor in the marketing mixture was the one thing that made the marketing plan succeed, just as it is nearly impossible to say what medium in the advertising campaign was the one that did the job. Many economists and marketing experts believe that marketing costs are unnecessarily high, but they don't know where to cut costs without endangering the entire marketing function. Yet, the need for finding a way to measure marketing costs has become apparent, too. Total marketing costs including wholesale and retail distribution often equal or exceed the entire cost of the raw materials and their processing and conversion into the finished products that leave the factories in search of customers. The trend is strongly toward increased efficiency in every aspect of marketing. What is happening, almost by chance and without conscious intention, is that the parts of the marketing function are being sorted out and examined, and that ingenuity, imagination, and research are being used to find ways to increase the efficiency of each part. The techniques of production are being applied to marketing, and the results are already seen in improved methods for many marketing activities, and most important of all in standards that are being devised to measure and check marketing's performance.

The trend to computers is well under way. The mathematical analysis and evaluation of every marketing activity is recognized as a practical possibility in 1964 and it is likely that in another thirty-five years this possibility will have become an accomplished fact. The pressure on marketing to find out what could be produced and sold at a profit has stimulated the marketing trends toward research and toward improved efficiency that point to a potential marketing capacity for finding new things to do that may again tax the abilities of production. In the past the cycle has always been from too much production and low prices to too much demand and high prices. Either the sales department or the production department was always in trouble because it wasn't able to fulfill its responsibility. Perhaps the most promising trend is that toward complete and accurate forecasting of the needs that remain to be satisfied. In every industry a beginning has been made in measuring the potential for its products in the next few years and in the more distant future. Realistic planning for the future is well begun in many enterprises and in some entire industries. Additional long-range planning has been undertaken by the government and by many academic institutions. Although it is true that much of this planning is fragmentary, disjointed, and unrelated to specific marketing objectives, in total it amounts to significant progress toward recognition of this essential marketing responsibility and toward development of the techniques required to solve it.

As long as marketing plans are made only by individual enterprises or even by industries, they cannot provide the assurance that is needed to stimulate additional production at a rate fast enough to keep up with the total available man hours!

What is encouraging is that this need for a complete evaluation of the potential markets for all of the goods that could be produced is being recognized and accepted by many truculent and independent business leaders. At the same time the research techniques for measuring potential markets are being tested and improved. Within another thirty years it is almost certain (if present trends continue) that a complete list will have been made of all the items that will be needed by all the different levels of the economy, together with estimates of the distribution of income within the economy that would assure the profitable production and distribution of various assortments of the additional goods that could be produced each year.

Such questions as these must be answered. If production were to increase by 20 percent, of what should that production consist? Where are the shortages? What are the needs? Or, if total production in 1999 were to be $2,000 billion instead of $1,700 billion, what additional goods should be produced, and to whom could they be distributed?

Better and more accurate research techniques are being invented every year and are being tried, tested, and perfected. The time will come, possibly within the next thirty-five years, when continuous market research will determine with strategically located meters, counters, tapes, and viewers exactly what consumers think of every product that is being produced. Like a daily census, the facts will be made known to everyone concerned. It will be possible for marketers all over the United States to tune in any of a myriad of clock dials that will be registering minute by minute the attitudes and actual purchases of a reliable sample of all the consumers in the country. It will be possible to note the changes in the supply and demand for any product from hour to hour, and to know what prices and promotions would be required to dispose of all the goods that might be produced. It will be clear that some items should be discontinued, and that others should be manufactured in larger quantities to sell at higher prices. The most suitable advertising appeals and the most efficient media will be determined and selected automatically by psychological consumer motivation studies that will be keyed electronically to the consumer research dials. As marketing problems are discovered, the memory of a computer will be probed to reveal their most effective and economical solutions.

As market research becomes more sensitive and more accurate in forecasting the total needs of the economy, it is clear that the uses of advertising will become more specific and more limited than they have been in the last fifteen to twenty years. The need for advertising that increased as the volume of goods to be distributed increased in the last decades will probably decrease again as the techniques for fitting the volume of production to the wants and desires of consumers are perfected. Advertising will no longer be used to test a potential market or to prospect for new customers. Research will already have determined how large the potential markets are, who the next customers are, and where they may be found. Advertising will have specific assignments to communicate the information that research has discovered and will motivate the consumers who are in need of and can buy the products to be advertised. The problems of creating pleasant associations and warm friendly *images* for specific brands will largely disappear when all brands together are produced in just the right quantities to equal the total potential demand. Although it is not the aim of market research to eliminate advertising, its effect will probably be in that direction. The effect of automated marketing in the U.S.A. will be to let the consumer decide what is to be produced. It will accomplish ahead of time an accurate measurement of the effective demand of an economy that will be somewhat larger than at the actual time of measurement due to the production of the additional goods that can be produced with an increasing supply of labor. It will be a continuous prognosis of the demand for goods that will exist if those goods are produced, dictated almost entirely by the wants and desires of the consumers.

The trend toward automated marketing is a trend toward rational and statistical competition. It is likely that a smaller number of larger producers will dominate most industries. As in the steel, soap, and soup industries today, competitive moves will be scrutinized thoroughly to make sure that they are economically sound, and when errors are made in packaging or pricing, they will be corrected immediately and almost automatically, not as a result of illegal agreements in restraint of trade, but as a result of the overwhelming pressure of competition between a few giant marketing organizations. Moreover, even the small manufacturer will have available all of the information he will need to recognize that an unsound competitive move would turn out to be ineffective and unprofitable. More wasteful and misleading competitive practices will become impractical as well as illegal as government controls become more numerous and inclusive. Better communication of all of the facts concerning an industry's sales, prices, and inventories, by markets and by brands and qualities of competitive products will also tend to lessen competition, as will the continuing trend toward standardization.

Competitive brands will tend to become nearly identical. They will be produced in many cases on identical machines and of the same raw materials, and in any case every brand will endeavor to match immediately any improvements that are discovered and are incorporated in any other brand. Differences in quality will tend to become more nearly equated with differences in price; and with automated production of large volumes of each item, costly changes in product quality will be less frequent and will spread promptly throughout an industry.

Price cutting will become ineffective and obsolete as a competitive tool. With prices equalized, and production costs very nearly the same among all producers, any variation in marketing costs will be reflected in an opposite variation in profits. Consequently, competition between the few large producers in each industry will tend to be for the largest possible profits at the price levels required to distribute the optimum volume of production. The optimum volume would be the volume that could be sold at a price that would result in the largest total profit for the industry as a whole. Price cutting would be futile, for in a stabilized industry it could lead only to inadequate profits, or to ineffective marketing which in turn would result in industry volume below the optimum volume for maximum profit.

Standardized products and prices will inevitably lead to increased standardization in selling and distributing the goods produced. As there are fewer brands, and general acceptability for all of them, the trend in advertising will continue in the direction of selling the product class rather than just the brand name. As in the tea industry today, advertising will have the objective of informing consumers of the product advantages shared by all brands, and will then undertake to condition potential users to think of a specific brand when they think of the product class. In a stable industry with a limited number of nearly identical brands, the share of market that will go to each brand will be determined by other factors than consumer acceptability. Competition will be greatest in the areas of service, availability, convenience in packaging, quality control, handling of product complaints, problems, and style, though it is very possible that most of these aspects of marketing will be standardized, too, within another thirty-five years.

Until recently Soviet economists gave little thought or time to marketing. Their problems were easy to understand and often impossible to solve. Their concern was to distribute the goods available as efficiently and equitably as possible. With all goods in short supply, the objective was to satisfy basic needs first and to allocate the remaining supplies where they would do the most good. The few comforts and minor luxuries available automatically went to the V.I.P.'s or were used as extra incentives and rewards for skilled workers and managers.

Until now marketing in the U.S.S.R. has been pretty much a matter of statistical allocations and logistics. There have been no problems of finding new markets for extra production or of developing techniques for persuading consumers to buy things they might do without.

Whether marketing will become as significant in the Soviet economy within another thirty-five years as it has become in the U.S.A. in the last thirty-five years will depend on many factors. It is certain that much more thought will be given to marketing problems and that new marketing skills will be required if the Soviet citizen is going to have a standard of living equal to that of the U.S.A. in 1964. In order to inform and educate potential users of the many new products that could be produced and that could enrich the lives of the Russian people it may even be necessary for the new Soviet marketing experts to rediscover advertising and sales promotion.

In both the U.S.A. and the U.S.S.R., marketing must guide the growth of the economy to assure the production of the goods and services that are needed and that can be distributed at a profit. Even if it should happen that before the end of the century the need for more goods is replaced by the need for education and art and time to appreciate and enjoy the beauties of the world, it will still be marketing's function to direct the total economic activities of the society to produce what is wanted.

It is for marketing to find out but not to decree. The total growth of the economy should follow the wishes of the people and not impose on them an abundance of goods that they don't want or need. This is why the marketplace is where their needs should be measured. What is worth producing is what can be sold for the full cost of its production.

Marketing skills may turn out to be as important tools for efficient economic growth in Soviet Russia as in the United States. Even more healthful, more pleasant, and more stimulating patterns of living may have to be learned.

As additional capacity to produce consumer goods is obtained, no doubt Soviet economists will plan to produce first of all the items that are in shortest supply and that are the most needed. Priority will go to increased supplies of food, clothing, and household equipment, which will be absorbed by the people of the U.S.S.R. with very little prodding. As in the U.S.A. in the 1930's it should not be difficult to find markets for substantial additional production of many basic products. However, the problem of increasing consumption from $548 per capita to $1,785 will require some assistance from marketing. The increase in the U.S.A. from 1929 to 1964 was accompanied by a dramatic increase in the use of advertising to inform consumers of the products that were available to improve their living standards. In the U.S.S.R. the problem will be to get from a level significantly below that of the U.S.A. in 1929, to a level only slightly under that of the U.S.A. in 1964; the problem will be to cover a much greater distance in the same length of time. An efficient system of communication will be needed, and it is likely that Soviet economists will continue to be interested in the marketing techniques that have been developed in the U.S.A.

Present trends in the U.S.S.R. do not point toward doing without any of the gadgets and frivolities critically associated with living in the U.S.A. Fancy clothes and furbelows are desired and needed in the U.S.S.R. just as much as anywhere else in the world. Maybe more! The interest of young people in the U.S.S.R. in Western hairdos and lipstick, in transistor radios and modern dance steps, tells more about Soviet desires and intentions than any official releases. And there is ample evidence from Soviet leaders that they are interested in modern supermarkets, department stores, and discount houses; that they are becoming as concerned with marketing as with production.

To raise Soviet living standards more than 300 percent in only thirty-five years will be especially difficult if housing and automobile production both continue to lag behind the rest of the economy. Even with advertising it may be impossible.

Of course, the U.S.S.R. has the option of not increasing living standards as much as the growth in the economy would permit. By a little careful trimming at each level, the total needs could be cut back by as much as 20 percent without changing the look of the market for consumer goods. Planned marketing techniques may be invented to tell each income group exactly what it is supposed to consume. Or rationing may be used to issue exact quantities of all the goods produced to those who are supposed to need them. An automated system of allocation and distribution is conceivable, although it does not look like a practical possibility.

Marketing's job in the U.S.S.R. is dictated by the production trends. It is expected that there will soon be enough of nearly all basic commodities. More and more comforts, conveniences, and luxuries will be available, but there will be a continued shortage of housing and automobiles for many years to come. Soviet marketing will have to learn how to distribute the necessities efficiently and equitably and will have to learn how to measure what goods will be needed.

Nothing is more important than a pair of boots to a man who has to be outside in the wintertime. Nothing may be less useful to the same man than a second pair of boots. Certainly it would be wasteful and sad to continue to produce more boots after enough have been produced to satisfy the needs of all the population. An extra pair of boots might be a good thing for most men to have, but if those men prefer to save their money to buy TV sets, that is what they will do. The workers might be better off with extra boots, but if they won't buy them, the economy might be better off to produce more TV sets!

Marketing men in the U.S.S.R. will have to consider what new markets there might be if the needs and desires of consumers were to be satisfied, and at the same time will have to think about creating markets for additional goods that the State thinks consumers should want. As in the U.S.A. the marketing men will have to balance the desires of the producers of goods against the desires of the consumers.

In a planned economy some hidden persuasion may be essential to convince consumers that they should buy the things they need instead of the things that they want! Propaganda and advertising will have to convince the people that they don't need motor cars.

Marketing may have to become oriented not to the consumer but more to the State and what it believes is best for the consumer to have, but even a planned economy will have to take into account the distribution of income and the ability of the people to purchase what can be produced. Marketing will have the job of measuring potential markets for additional production by income groups and by areas, and will have the responsibility for informing the State what goods could be marketed and at what cost to the economy.

Consumer research in the U.S.S.R. should be relatively easy to do. No permissions will be needed to get pantry shelf inventories and audits of current purchases. The families selected as respondents to be interviewed will be more than willing to tell all they know about their buying habits and intentions. It should be possible to get detailed data concerning everything that is used by each segment of the population in every section of the country. No doubt a statistical record will be put together showing exactly how many of what goods would be needed to raise the living standards of any part of the population by any number of percentage points.

When the State has determined what goods will be produced and what potential markets are to be satisfied, marketing will then have the more delicate assignment of allocating and distributing the additional goods produced to the people for whom they are intended.

Perhaps this objective will be easier to accomplish in the U.S.S.R. where prices can be adjusted to make goods available to those who need them without regard for profitability.

However, the marketing man in the U.S.S.R. will find that he cannot escape all responsibility for profits. If he has to sell goods below cost in order to get them consumed, he will have to get an extra price for something else to make sure that the total purchasing power is used up in buying the goods that are available for sale to consumers.

In the U.S.S.R. the amount available for investment in new production will be equal to the difference between the total wages paid after taxes and the value of the total production achieved. In both cases the capital investment is equal to the total profit of the economy, if the profit is assumed to be the value received in work less the amount paid out to the workers in the form of consumer goods.

It is misleading to talk about distributing profits. What is distributed is a claim against future production, a claim for repayment of the work invested, the work that was used to produce capital goods. In the U.S.A. this claim is represented by stocks and bonds and notes, and the work given up is assumed to be the work or the proportion of earnings that was not spent for consumer goods by those who had surplus incomes. It is assumed that the extra work invested was done by those who have earned surplus incomes. Thus, the claim against the future is distributed among those with surplus earning power in the present.

In the U.S.S.R. no recognition is given to this claim for repayment of the work invested in the future. It is assumed that whatever increased production is achieved in the future will be used in accordance with the long-range objectives of the State to improve living standards for everyone and to move as rapidly as possible through socialism to communism.

The principal difference is that no one in the U.S.S.R. can get a vested interest in future production. Some individuals may get a larger share, a much larger share, of current production than others, but the share they get is redetermined regularly, and is based on the contributions they are credited with making to the State. Thus, in the U.S.S.R. the total value of the increased production remains at the disposition of the State and is distributed in accordance with a plan that takes into account the needs of various segments of the population and the rewards and incentives earned by each.

In the U.S.S.R. all the profits are reinvested without the consent or approval of the owners, and the value of each share in the total economy is increased proportionately. Whether the people, who are the State, are the owners is debatable. As in every other state, powerful individuals and groups of people manipulate the policies and precepts of the Soviet economy, and to the extent that the political and social structure of the State permits, they can be expected to rewrite the rules to suit their own ideas and ambitions.

Compared to the hand-posted ledgers and accounts of only a few decades ago, the systems and machines available in 1964 for collecting and recording statistical facts can provide a hundred times more information concerning any aspect of a business than anyone in the 1930's would have imagined could be of any use, and can produce all the facts in a succinct and organized pattern with incredible speed and accuracy. Another thirty-five years of growth and change at the same pace hardly seems possible. Yet the pace has been increasing every year. The possibilities of today's electronic computers are only beginning to be recognized, and yet the chances are good that these computers will have become obsolete and will have been replaced by new devices not yet conceived in less than thirty-five years.

By that time it is reasonable to expect that the relevant data about the economy will have been committed to memory by some kind of machine, and will be in the record. The job is already begun. The facts are going into the machines at an ever increasing rate, and the cost of gathering and processing an always larger and more comprehensive volume of information is turning out to be less than it used to be to maintain relatively simple records by hand.

Not only will there soon be a record of the growth and experience of every business enterprise and every industry but there will also be a record of the remaining resources of raw materials in the country and in the world, and an accurate estimate of the cost of making them available wherever they might be needed. The statistical facts about the total supply of labor, classified by skills, age, character, geographical location, and present earnings will be stored permanently in the system together with similar statistics concerning every corporate and individual enterprise in the country.

If there's not much that isn't known now about every unit, human or machine, in the U.S.A., whether we like it or not, there will be much less that isn't known in another thirty-five years. Not only will the facts be recorded, but the future status of each unit will undoubtedly be estimated, so that the effect of any shift in the direction of the economy will be measured and balanced immediately by means of adjustments that will be weighed, tested, and selected electronically.

It must also be assumed that the information available will be used objectively in making plans for the economy. The trend in this direction has led already from the suspicious and reluctant acceptance of some theoretical suggestions offered by the economists in the 1930's to a continuous stream of scientific seminars for businessmen in the 1960's. The courses sponsored by every kind of business association in the U.S.A., from the A.M.A. to the Chamber of Commerce and the books and periodicals dealing with techniques of management, finance, production, and marketing are ample evidence of a trend that is just beginning.

In as short a time as another thirty-five years, it is more than possible that there will be fewer seminars and crash programs, and that a mature business management will be operating its industrial units and the entire economy with confidence in the facts, procedures, and techniques that govern its decisions. Even now, a few industries are already operating with scientific precision and assurance.

If this trend continues for another thirty-five years, the idea that any enterprise could be operated unscientifically, without access to and regard for all the information that could affect its success, will be inconceivable. Facts will be at hand, available to the small merchant as well as to the international corporation. Data will be recorded anywhere in the world, and will be instantaneously available everywhere else in the world. There will be tools available to select what is relevant and significant for answering any problem, and tools to draw the right conclusions from the facts. The man who wants to open a small retail store will punch a card with the data about himself and his proposed enterprise, and will get back in a few minutes a forecast of the results that he can expect from such an enterprise over the succeeding five years. And the large corporation will not only operate like a machine, but will virtually be a machine! Nearly all of its decisions will be taken automatically, by referring to a procedure and consulting the records and measuring projected results of possible alternatives. How can decisions be taken on any other basis when the records are available and the results of any conceivable action can be projected with reasonable accuracy? No executive's personal experience could possibly include all that is contained in the computer's memory, and no executive could remember all that he has experienced as accurately and completely as can the computer. Nor is the executive's intuition sharp enough to perceive all of the combinations of factors that can be measured by an electronic machine, when it is programmed to test the small and subtle differences that relate to any single specific problem.

The trend in financial skills is as dramatic and as significant as the increases in debts and government expenditures. Today the financial experts not only know in accurate decimals exactly what has happened and how much it costs in every production and marketing department, but they know what is happening now, almost every minute as it happens, and they can predict, and do, what is going to happen months ahead of time.

But that's not all. They not only predict what will happen, they point out what must happen and where and to whom if various alternative financial objectives are to be achieved. Their skills and their techniques have made it possible for them to ensure the continuous profitable operation of large and complex industrial enterprises, and to recognize and diagnose the symptoms of failure, and even to prescribe the surgery or other corrective measures that may save the ailing business.

What has created electric calculators and computers has been the need for faster and better tools to get the facts safely recorded, and to provide all the kinds of information wanted by the tax collectors, the payroll auditors, and the forward-planners. And the computers have made possible the accumulation and presentation of more and different facts.

So the progression seems to go on without end. In a philosophic mood it might be called an example of mankind's thirst for truth. In the field of finance, business has progressed from the use of a very few facts, but about all it could afford to have noted and written down with a quill pen in a ledger, to typewriters and adding machines and a need for more detailed information, to automatic calculators and billing machines and still a greater need, and on and on—always to better machines and better methods for getting the facts, and never any end to the additional information that could be used.

The trend is toward a battery of computers that can control within a very small margin of error every cost of production and every activity of marketing to guarantee a continuous balance between the revenue produced by marketing and the costs of production and distribution. The rate of growth of every corporation will be regulated exactly and finally by the cash flow and surplus earnings that it can produce.

The financial management of a corporation can thus be formalized, and programs for various alternative courses of future development can be prepared and entrusted to an electronic brain. Even the activities involved in raising new funds and selling stock are becoming routine mechanized operations whose results can be guaranteed within a fraction of a percent.

Automated finance can be expected to eliminate the problem of security altogether within thirty-five years. There can be no question of security if an enterprise is assured of success in advance by being able to know what quantitites and kinds of goods it can sell, and what prices it can charge, what the costs of production and marketing will be, and what it will have to pay for money. This is not the place to look ahead more than thirty-five years, but it is intriguing to imagine what will happen when there is no more speculation. Much so-called speculation in the 1960's is less speculative than it seems. Growth is planned as an alternative, and a profitable one, to higher dividends that could be paid, or to lower prices.

A continuation of the present trends could produce a stabilized financial economy within another thirty-five years, with funds flowing automatically to the enterprises where they would be most effective and profitable. The need for such a stabilized financial atmosphere is seen in the additional trend toward larger and more far-reaching enterprises with activities extending around the world. Concern with growth as opposed to immediate income is another essential for financing the long-term investments in new frontiers, in Africa, India, and South America, that will be required to maintain the industrial leadership of the U.S.A. The financial trends seem to be leading in that direction, toward efficient, mechanized financial controls that will create a stabilized financial market serving the entire world.

What communication alone can do in the sober field of economics is rarely appreciated. Communication can promote recognition, understanding, even sympathy, and make possible an exchange of experiences and ideas. It can extend knowledge and stimulate new ventures through emulation and competition. It can stir emotional reactions of fear, envy, pride, pleasure, and love. It can reveal similarities in industrial and corporate patterns of activity, point up differences, and raise questions. By so doing, it can hasten conclusions and answers.

Communication is a force for enlarging economic knowledge and understanding, and for encouraging industry, trade, and cooperation. It is a potent force for peace, good will, and cultural growth—for a healthy world economy. In the U.S.A. the ability to deliver information swiftly and abundantly to every manager and every worker will prove to be a vital factor in determining the economic growth that may be achieved in the next thirty-five years.

If these economic trends continue for another thirty-five years, it is possible to imagine that the economy of the U.S.A. will be controlled, operated, and planned by machines. Multitudes of mechanical monsters will gobble up all of the facts and spew out the blueprints for every kind of automated activity from the harvesting of corn and the brewing of beer to punch-card-controlled assembly lines and push-button supermarkets.

Even purchasing a new car will be automatic, with a card punched for the trade-in and a record of every available model in town. The buyer will mark a card with the statistics about his old car and will feed the card into a slot under the net price he is able to pay. Out will come a set of photographs and descriptions of all the cars available that suit his requirements. He will then make his choice, attach his down payment to the photo, and go to the delivery garage where his old car will be ingested by an automatic parking device and his new car will slide down the ejection ramp untouched by human hands.

With automated production, mechanized marketing, punch-card finance, and electronic administration the economy will function like a mathematical formula, adjusting continuously to the information concerning need and wants, production techniques, market research, hungry children in Africa, war threats in South America, new machines in Russia, schools and roads, manpower and capital resources, birth rates, and depreciation.

These are the trends and they all appear to lead in the same direction. This will be an economy that is hypnotized and fascinated by statistics, and that is moving at an accelerating pace toward the goals of optimum production with maximum efficient use of all the raw materials available and all the power, energy, and imagination that can be mustered from outer space or inner man. This kind of machine cannot be controlled by a board chairman, a business association, a labor union, or a state or federal government. It is purely pragmatic. Yet, its objectives are *peace and prosperity* for it will be a dynamic and optimistic economy that seeks continuous growth and improvement and one that will be sparked by human beings who, in turn, will be continuously healthier and wiser than their predecessors.

19

Scientific Management

IN Soviet Russia, probably too much scientific management was applied too soon and in too arbitrary a fashion. A political bureaucracy was created to run the country, and an economic bureaucracy grew up alongside it to dictate every move of every economic enterprise. An attempt was made to reach all decisions objectively and rationally before there was enough information available to support such decisions.

Nevertheless, a beginning was made thirty-five or more years ago, and in that time the Soviet economists have gained considerable experience and practice in applying their theories to a rapidly expanding industrial economy. Whether they have learned or are learning fast enough to keep up with the increasingly complex problems they will face in the next thirty-five years remains to be seen. It appears that they are putting together the machinery and the organization that will be needed. More and more statistical data is becoming available, and if present trends continue, by 1999 every possible fragment of knowledge that could be remotely useful will have been gathered and stored in some kind of electronic filing cabinet.

Questions concerning the accuracy of Soviet statistics have been raised in the past, partly because of the inexperience and lack of training of many of the people charged with putting down the figures in the first place, partly because of the pressures on many production units to falsify or magnify their achievements and to minimize their costs, and partly because of the lack of suitable modern equipment for communicating, recording, and analyzing statistical information. These doubts will diminish as the tools for measuring and recording the raw data improve. Better mechanical equipment in the form of calculators, tape recorders, teletypewriters, and computers operated by trained accountants and highly skilled data processors will soon assure the accuracy and validity of most Soviet statistics.

Soviet management is also improving its ability to use the available facts intelligently. With the aid of new digital computers and analyzers it may be expected to keep up with the ever increasing load of factual information. If the present trends continue, it is possible that the programming of electronic computers may soon compete with chess as a mark of intellectual superiority in the U.S.S.R.

Scientific management needs a plan, or an objective, that prescribes what needs are most important and in what order they should be satisfied. As yet, no statistics have been discovered that provide a scientific answer to the question of what needs should come first. New factories, nuclear rockets, homes for workers, food and clothing are all needed. Which are most important? This is a question of values, morals, philosophy, and politics. In the U.S.S.R. the answer is dictated by the State, and political considerations weigh heavily in the balance against health, productivity, and popular demand.

In the U.S.A. the answer is dictated largely by price, by

what people, including corporations and governments, will pay for what could be produced. In the U.S.S.R. it is assumed that what consumers will be permitted to purchase can be controlled by setting prices arbitrarily to absorb the funds on hand that could be used for consumer goods. The State decides how much of total production will be allocated to defense and other costs of government, investment in industry, and consumer goods. The Soviet economists are then supposed to determine how much of what goods should be produced to satisfy the needs of industry and of consumers. This determination must be made without a valid record of costs and prices inasmuch as costs are arbitrarily set by controls on wages, and prices are related only to available supplies and what the traffic will bear. It is possible that research can be devised to find out what is needed most, and how much should be produced of every item that can be manufactured in the Soviet Union. It would seem that such research will be required to complete the job of scientific management in the U.S.S.R. In the meantime it is inevitable that many decisions with regard to what will be produced will be made arbitrarily.

In 1964, the U.S.S.R. is about as far along the road to scientific management as was the U.S.A. in 1933. It should also be noted that if the Soviet Union is as interested in managing its economy scientifically as it has been in the development of nuclear power, it is fairly certain that errors and inaccuracies will be corrected when they are discovered. If it should turn out that valid prices and costs are needed to allocate production efficiently, it is altogether possible that the laws of supply and demand will again be invoked whether or not they conflict with Communist sentiments. If it should turn out when all of the data have been collected and digested that some basic theories of communism are in error, it is more than likely that those theories will be abandoned in practice, if not in theory. Some such adjustments in Communist thinking have already been made and the extension of scientific and mechanistic controls throughout the economy may bring about more changes. It is possible that the U.S.S.R. believes in science even more than it believes in communism.

Whatever may happen to Communist economic theories, a continuation of the current trends in the U.S.S.R. for another thirty-five years will produce an economy much like that predicted for the U.S.A. All of the information will be available on punch cards, the number of pounds of beans, pork, carrots, and coffee consumed per capita each week of the year in each village and city in each of the Soviet republics, the exact number of neckties and nightgowns, of tires and typewriters wanted in every section of the country, the quantities of pipe and petroleum and the kinds and numbers of machine tools, trucks, and railway cars required for various economic objectives will all have been calculated and checked, reviewed and recorded. The potential production of every productive unit in the U.S.S.R. will have been carefully estimated, under all the possible circumstances that could affect its productivity, including weather, quality of raw materials, age and condition of equipment, and the skills of the available personnel.

What could be produced and where and how will be known, and will be related to the potential supplies of power and labor. Production plans will be contained in computers, and will be adjusted automatically to cope with changes in the products required and the raw materials available.

An equally thorough, impersonal, and complete inventory will be maintained of the skills and potential productivity of every individual human being in the U.S.S.R. Punch cards will tell all there is to know about every worker, every machine, and every factory.

Despite the similarities that are developing in patterns of management, there are striking differences in the point of view of the two economies. Management in Soviet Russia embraced the idea of science long ago, but was held back by the lack of information and trained personnel. Managers in the U.S.A. have been skeptical of scientific management for their economy until quite recently, but have proceeded just the same to gather information and to use it in making decisions that have become more and more scientific. The U.S.S.R. is committed to proving that a command economy can operate scientifically. The U.S.A. is committed to proving that a market economy is more efficient than any other economy whether or not it is scientific.

In fact the U.S.S.R. is moving toward what will be an acceptable replica of a market economy whether or not it succeeds in substituting consumer research for prices in determining needs, and the U.S.A. is moving rapidly toward a managed economy even though it may retain a free marketplace. Today in the U.S.S.R. "the division of the national income into investment and consumption is actually made in the physical plan" for what is to be produced.[1] When physical plans for production based on the information obtained from computers are made in the U.S.A., a similar allocation of national income to investment and consumption will be made. When and if this will ever happen depends entirely on whether the facts will ever be available that will tell intelligent managers how much of what products could be produced and distributed efficiently (at a profit). When the facts are available scientific managers will use them in much the same way in the U.S.S.R. as in the U.S.A. The main difference will be in how the objectives are set. In the U.S.A. they will be set by the market, in theory, subject to some necessary controls by the State. In the U.S.S.R. they will be set by the State, in theory, subject to some inescapable pressures from the market.

[1] Robert W. Campbell, *Soviet Economic Power* (Boston: Houghton Mifflin, 1960), p. 148.

20

Meaning of Trends in Typical Communities

KRIMSK and Kalamazoo will change again before the end of the century, as much as or more than they changed from 1929 to 1964. Kalamazoo is expected to double in size, from 183,000 to 375,000, and Krimsk will more than double, from 30,000 to perhaps 80,000, as the Krasnodar area increases from 450,000 to nearly a million.

Incomes and living standards will increase, too. In Kalamazoo the number of people in families with incomes of more than $10,000 will increase from 36,000 in 1964 to about 225,000 in 1999, from 20 percent to 60 percent of the population. More than six times as many families will have more than $10,000 and of these one-third will have more than $30,000—all in constant, uninflated 1954 dollars. No family in Kalamazoo will have less than $5,000.

In Krimsk in 1964, only 5 percent of the people, 1,500 in all, have family incomes in excess of $5,000. None have incomes as high as $10,000. In 1999, 10 percent will have family incomes in excess of $10,000, and half will have more than $5,000. The number with incomes of $5,000 or more will increase from 1,500 to 40,000. In 1964, 95 percent of the people have family incomes of $2,000 or less. No one in 1999 will have living standards as modest as nearly everyone has in 1964. The poorest 10 percent will have family incomes in excess of $3,000.

In 1999, Kalamazoo will be known for its colleges and cultural activities as well as for its industry. With the work week down to thirty-five hours, there will be time for reading, music, drama, and sports. Leisure, plus education and wealth, will increase the interest in painting and poetry by 1,000 percent, and if golf only holds its own, the seven golf courses in 1964 will have to be increased to more than forty in 1999. There will also be proportionately more bowling, tennis, ice skating, and bridge. There will be more and better retail stores, too, for the 60 percent of the people who will be able to live in and furnish homes like those of the top 20 percent in 1964.

Kalamazoo will continue to spread into the country, and throughways will lead from the center of town to new suburban communities with their own shopping and service plazas. For an increasing number who don't care about gardening and grass cutting, there will surely be luxurious garden apartments with private patios and swimming pools near the civic auditorium and the combined institutes of art that will naturally grow out of the 1964 art center. The acres of old houses near the center of town will be gone, replaced by new office buildings and department stores, a heliport, and an imposing collection of exclusive shops. Although it will be possible to go to Chicago or Detroit in an hour by plane, it will be equally possible to come to Kalamazoo, and branches of many internationally famous companies will now be found around the corner from the Park Club, if perchance it is allowed to remain as a landmark from the past.

In 1999 Krimsk will be a thriving industrial city. Modern factories of steel and glass will be built on a wide industrial parkway that will extend a mile or more in either direction from a new *combinat* office building, connecting at each end with the turnpike from Krasnodar to Novorossisk. In addition to the original canning factory, now turning out more than double its 1964 production, there will be an automated frozen-food plant processing frozen fruits and vegetables, TV dinners, apple pies, and frozen borsch. Another new *Avtomat* will make plastic containers, and others will package baby foods, breakfast cereals, and drugs, and will manufacture refrigerators, detergents, tires, and sporting goods.

Let's imagine how Krimsk will look in 1999. The busy freeway from Krasnodar has a curving exit that swings over the industrial parkway and up the hill into town without a pause. Although there aren't any tolls to pay, there is a control station through which all cars must pass. Trucks and other commercial vehicles carry IBM cards on which the time of arrival and departure is punched, and other than local travelers must still exhibit their travel permits. Anyone can travel as he wishes but his whereabouts are always known in case he is needed.

The main street is all new since 1964. There are many attractive shops with ample supplies of goods produced in the U.S.S.R., and a few delicacies from elsewhere in the world. The Univermag (department store) on the corner is half a block long and four stories high and has everything for sale from TV sets to paper towels. Next door on the street that goes down the hill to the *combinat* is a spacious outdoor café, and across the way there is a new movie theater on the corner, next to the Lido Restaurant that specializes in French and Italian dishes and fish from the Black Sea. Farther along another gay dining room and cocktail lounge competes for the favor of visitors from all over the world. Where there was a simple bus stop at the beginning of the wide main street there is now a glass-enclosed terminal and garage. Buses leave frequently for Novorossisk, Krasnodar, and Odessa, and there is also a supply of cars that can be rented by the day or week.

In the other direction, the boulevard turns on up the hill past blocks of apartments with pleasant balconies and air-conditioned solariums, and then to the newest residential section where there are several hundred ranch-style homes, and an equal number of two-story cooperative apartment buildings with large enclosed gardens and orchards. In 1999, everyone in Krimsk has an apartment or house with hot and cold running water. Every person has at least fifteen square meters of living space.

The boulevard ends in a large circle where there is a complete shopping center with its supermarket, laundry, and automatic bowling alley pleasantly set in a park of trees and flowers. It all looks much like parts of Kalamazoo except that little space is provided for private cars. There are still few of these in Krimsk. Bicycles, motor bikes, and shopping carts are lined up around the circle of stores and a small section in back of the buildings is reserved for trucks and a few larger vehicles. Many shoppers use the convenient local buses that give frequent service to every part of town, and it is now possible to have purchases delivered at a small extra charge.

Past the circle in a modest campus are found the buildings of the Institute for Food Technology, and a little farther out at the end of a country lane there is the Krimsk Golf Club where for the last ten years teams of young Soviet athletes have been learning to putt straighter and hit the ball farther than the capitalist pros.

All of this is new since 1964, as is the high school that can be seen in the distance as we go back down the hill toward the *combinat*. In between are many of the original apartment buildings with their trees thirty-five years older but otherwise much the same. The feeling of Krimsk hasn't changed greatly, but its industrial growth is reflected in its increased living standards and new interests.

In 1964, there was literally one class in Krimsk. The director of the *combinat*, a lean, square-jawed, determined young man, the laboratory technicians in white aprons, the engineer who knew all about the taste problems involved in putting beer in cans, the schoolteachers, truck drivers, and manual laborers in the factory, the children in school uniforms, and the old men with droopy beards—all seemed to belong to one family. Peasants, proletariat, or workers, no matter what their income or education, they were all equally important if not exactly equal in status. In 1999, there are greater and more significant differences.

There is now a technical institute with a staff of professors and scientists. There are more professional people, doctors, economists, and accountants. There are more managers and directors earning more money and living a little better than anyone else. There may still be one class, but its unity isn't as apparent as in 1964. Then, only a very few, if any, department heads and engineers earned as much as $600. Perforce they spent most of their time in activities that included workers from every part of the small economy that was Krimsk. In sports, and school celebrations, and community planning all lived and worked together. In 1999, the higher income group is much larger. When only a few families in a community are even moderately well off, a one-class society is natural and easy to maintain. In 1999, it is not so easy to avoid some class distinctions. There are too many educated and sensitive people who are able to demonstrate their individual tastes and ideas in their way of living. Those with similar incomes and similar tastes tend to do things together and to pursue the same hobbies and interests.

The road from Krasnodar to Novorossisk is now a 70-mile-an-hour throughway and Krimsk finds itself conveniently situated to send its many new products all over the world in ships that sail from Novorossisk to India and France and the U.S.A. Supplies including machines and tools from the United States, flow in from everywhere on nuclear-powered ocean barges. A fleet of trucks takes its production, mostly processed foods of one kind or another, to markets as far away as Moscow and Leningrad on the straight and semiautomatic highways that now connect Vienna and all of Western Europe with the U.S.S.R.

Although Krimsk has a local airport for small planes and helicopters, its outlet to the rest of the world by supersonic jet is through the bustling city of Krasnodar, now a metropolis of over a million people. The people of Krimsk do have access to the rest of the world in 1999. They can and do fly to New York nonstop in just a few hours. Many young people have studied in the United States, and there are students from Kalamazoo College at the Institute for Food Technology in Krimsk. Since 1975 when travel outside the U.S.S.R. became possible, at least a tenth of the people have visited some foreign country on business or for a vacation, and nearly all have returned.

In 1929, less than a third of the people in Krimskaya could read or write; in 1964, literacy reached a level of 97 percent and most young people received the equivalent of a high school education plus adequate technical training; and in 1999, almost no one under the age of forty has had less than two years in a university or a thorough academic background in his own special technique. A large number, including most managers and department heads, have completed what amounts to graduate college work in their fields. There are a few who have devoted their lives to history, literature, or the fine arts.

In Krimsk there are now two legitimate theaters and several popular movies where films from all over the world are presented. There is a symphony orchestra that gives local performances throughout the year and makes a short tour through the south every winter, and there are small dance bands affiliated with each of the larger factories. In the park near the *combinat,* an open-air pavilion for dancing that has been popular for many years has recently been provided with a removable cover and a heating system for use in the winter. There is also a pleasant restaurant in the new shopping center near the Institute which offers music and dancing two nights each week. However, most people go to Krasnodar for evening entertainment. It is less than an hour away on the throughway, and Krasnodar has the latest films, and internationally famous singers, dancers, and players with the latest productions from Broadway as well as the classic ballets from Moscow. In Krasnodar, there are now restaurants and nightclubs as well-known as those in Paris and London.

Most people in Krimsk still don't have their own personal cars. Out of 25,000 families, not more than 4,000 have even one car. However, everyone has a car to use when he needs it. The *combinat* rental service is inexpensive, and now has a good supply of cars. Until recently it was necessary to reserve a car several weeks in advance, but lately it is almost always possible to get a car on less than a day's notice. And the new buses in Krimsk are fast and comfortable. There is good service to Krasnodar and Novorossisk; to go to Anapa on the Black Sea for a swim and an afternoon in the sun takes hardly an hour each way, and costs less than a dollar.

All the food-processing plants in Krimsk are now completely mechanized, and have been for many years. The last hand operations were eliminated some time in the 1970's. The plastics factory, and the frozen-food packaging operations are automated from one end to the other. With only two and one-half times as many workers as in 1964, production in Krimsk has increased by more than five times, and many new projects are being planned that will be begun as soon as the necessary manpower can be found.

Perhaps the manpower that will be needed can be found in Kalamazoo in 1999. In a completely automated industry the trained technicians and programmers will be available on a lend-lease basis to help set up and operate new projects anywhere else in the world. Although everyone will be employed in Kalamazoo in the 1980's and thereafter, there will always be extra hours that could be worked when a team of specialists is detached from a local industry for service elsewhere. By 1999 the total world production of paper, pills, and peppermint will be pretty well-known through reliable statistical information available in Kalamazoo and Krimsk alike, and it will be easy to determine what additional needs there are and where new productive facilities should be located.

Perhaps automation will contribute to the softening of class distinctions in Kalamazoo in 1999. Increased productivity and an increase of more than 100 percent in average incomes will change the structure of society. When six out of ten families can live in the finest homes and travel frequently to Paris or Rome, and when anyone can afford vintage wines and sirloin steaks, the differences that there are in incomes will become less significant. And when almost everyone has had a college education, differences in interests and tastes will be less sharp and more a matter of degree and emphasis than of basic attitudes toward life. In Kalamazoo there won't be anyone left in 1999 to occupy the many old-fashioned two-story frame houses that fill large sections of the city in 1964. Most of the old houses will be gone and many newer ones, too.

In 1963 in Krasnodar I asked my guide to show me the poorest part of town, and she didn't know what to do. There wasn't any poorest part or any best part. It was all the same. We did drive from one end of town to the other and did find one house that dated back to the 1790's, and a number of cottages without any modern conveniences, but we also found the same new apartment buildings everywhere. There wasn't any very good section or any that was poverty stricken.

In Kalamazoo it was easy enough to locate the poorest section of town in 1964, though, in fact, it wasn't too poor by

Soviet standards. In 1999 it won't be as easy. A few relics will still be scattered through the city, but almost everyone will want and be able to rent or buy a small and trim modern home or a suite in an apartment with a swimming pool.

The cheap taverns and shabby cutrate stores and pool halls will be gone, too. There weren't too many in 1964, and they will be forgotten in 1999, as will the crowded and ugly trailer parks near the edge of town. These are changes that seem inevitable with increased leisure, more education, and actual affluence. Even if the growth rate is only 2.5 percent annually, this is what must happen. Kalamazoo will become a conservative, friendly Utopia. Even the sober square downtown will be spruced up and embellished with abstract sculpture, and shining pools, and terraced walks. All of this and more will happen if there is peace and if the opportunities for economic growth are granted even a conservative welcome. Whether there are forty golf courses or forty glider fields, there will be sports activities going on all day and under the lights at night, and there will surely be golf and bastketball and hockey teams from Krimsk competing in Kalamazoo.

Krimsk and Kalamazoo will be only a few hours apart in 1999. They will know each other, and will exchange ideas and understanding. Perhaps they won't believe all the same things and won't worship the same gods, but more than likely they will work with the same tools, play the same games, and have many similar economic goals.

Their economies will still be decades apart. Krimsk will do well to catch up another fifteen years in GNP and per capita income, but the chances are good that Krimsk will at least share the lead in science, sports, and the arts.

Together, Krimsk and Kalamazoo in 1999 will disclose the prospects for the century ahead. Their economic achievements will show how far economic growth can go in satisfying man's needs, and their patterns of living will suggest how their civilizations will continue to develop.

21

Summary and Conclusions

BRIEFLY, my conclusions with regard to the prospects for the U.S.A. and the U.S.S.R. are that:

1. the potential growth of each economy is not inhibited by the parallel growth of the other,
2. these rates of economic growth will not be determined by their political philosophies,
3. the actual growth achieved will have an effect on their political philosophies,
4. the projected growth does depend on continued peace in the world, and
5. the present trends do lead in the direction of world peace.

The economic trends in the U.S.A. and the U.S.S.R. since 1929 weren't legislated by Congress or decided by management committees. They may have been planned, as in the

U.S.S.R. but in the end they were determined by the needs and possibilities of the two economies. The trends can be modified by legislation and by the attitudes of labor and management, but their direction and potential is limited by the stage of growth of the economy and by the quantity and quality of the available capital goods and the work force.

This means that economic trends are not capitalistic or communistic. They are economic. They may be affected by the political philosophy of the State, but their direction must follow the availability of capital and labor. What happens is that they reflect changes in the economic situation, with adjustments in living standards and the development of new needs and different allocations of productive capacity, and changes in the economic situation affect and change political philosophies. The projected changes in the economic situation of the U.S.S.R. will surely bring about further changes in the political philosophy of Soviet Russia. How the U.S.S.R. will look in another thirty-five years will depend on the growth that has been accomplished, the standards of living that are being enjoyed, and the success that has been achieved in stimulating and making effective use of the potential talents and skills of all the people.

Socialism and communism weren't created in a vacuum—nor was capitalism. They were the product of many pressures —social, political, and economic—that became effective when economic growth and change created a favorable economic situation. The facts of economic life can be measured in their social and political effects as well as in the volume and variety of the goods produced.

That there will be further changes in the political philosophies of the U.S.A. and the U.S.S.R. as a result of further economic growth and change is more than just a possibility. It is inevitable.

A visitor to the U.S.S.R. can almost feel the changes that are imminent in the attitude of the State toward communication and contact with the U.S.A. Guides and directors of farms and factories everywhere are pleased to have American tourists and businessmen come to see their activities and invariably remark that there have been many others from the United States this year. And when they are invited to come to the U.S.A. for a visit in turn, they are wounded and hurt because they know they cannot, and know that this is wrong. A few years ago they weren't hurt. They just shrugged their shoulders. Such an idea was ridiculous. Who would want to visit the enemy? Now they care, even though some have visited the U.S.A. in the last few years, and more are planning to come. Those who have had a chance to visit any parts of the world outside the U.S.S.R. make it a point to mention their travels in even the briefest conversations.

In 1964 it is almost certain that if everyone were free to go and come as he liked, many would leave for the U.S.A. and other places with the intention of staying. Some older people would leave because they still harbor fears and resentments that can never be forgotten, but many younger ones would be attracted by the glitter and promise of America. The number would probably be a small percentage of the total population, a fraction of 1 percent, but perhaps as many as 300,000 in all. If anyone could go, then each would have to weigh the opportunities and risks for himself. He would measure better housing and more expensive living in the U.S.A. against a safe and sure average standard of living in the U.S.S.R. Only the more adventurous would dare to go, and many of them would hesitate. Many would come back after sampling the advantages and disadvantages of capitalist life.

But in 1964 the risk is still too great for Soviet Russia to take. Until there is ample food and other consumer goods, a shorter waiting list for cars, and better housing, the restrictions on travel cannot be lifted. Another twenty years, at most, will be needed, maybe less. By then there will be few left in Russia who haven't been brought up entirely in the

Soviet way. And by then the Soviet standards will be at least twice their levels in 1964. Even by 1975 there is a realistic chance that nearly all Russians will be free to go and come as they choose, after arranging for the time off, and after saving the necessary money. Exchange students, scientists, and technicians will lead the way, and more and more others will have a chance to follow.

This change in attitude reflects the changes in the Soviet economy that have been accomplished already, and it is an example of the kind of changes that may be expected in the future. Economic changes involve changes in the people who live and work in the economy, and who have new standards with which to measure the necessity and significance of the many activities that make up their daily lives. And, as the people change what they are doing and the way they are living, they change their attitudes toward wealth and poverty, work and international trade, the distribution of income, and wage rates and profits.

This argument is not based on the assumption that the leopard can or should change his spots. The argument is simply that his spots will change when the economic situation in which he lives changes.

The Communist party line will change when living standards in Soviet Russia are as high as those in the U.S.A. in 1964, and thinking in the U.S.A. will change as poverty and deprivation are eliminated, and an economy of abundance becomes a fact.

Ideologies have a tendency to be believed and quoted as gospel long after the economic situations to which they are related have completely changed. Ideologies that remain intellectually defensible and emotionally acceptable are hard to kill. That the facts change and the reason wears thin makes little difference. In the absence of new theories that combine evident validity with the common touch, the old familiar ideas still hold on to a large share of the public mind even though they may be forlornly out of date and inappropriate. Communism and capitalism may be equally unrealistic in relation to the economic situation as it can be measured in 1964. Neither may fit the current economic trends, and the modern techniques of production and distribution, but at present no suitable substitutes are available.

A theory that takes account of full employment, automated production, comfortable minimum standards of living, suitable rewards, and incentives for creative achievements and hard work, high civic and family morale, international understanding and good will, free world trade, sophisticated systems for sharing nuclear power, and all the other potentials of modern science is hard to condense into a phrase as simple as "from each according to his ability and to each according to his need," or as succinct as "the profit system." Even if it includes new definitions of *profits* and *needs* that enable it to encompass both communism and capitalism it may have difficulty in gaining any supporters from the adherents of those fighting philosophies, and even if it is entirely scientific it may have trouble finding any believers unless its logic is concise and clear.

What is needed is an ideology that contains the gist of the economic progress of the last fifty years and the sum of economic research and knowledge. It is not certain that such an ideology will ever be discovered but whether it is or not the facts won't change. Both the Soviet and the American economies will continue to change, and they will leave their mark on people's lives and ideas. Economic changes will continue to affect the values and ideals of individual men and of societies.

A new ideology that could capture the meaning and potential significance of the economic situation in the world in 1964 would have its effect in turning men's thoughts away from communism and capitalism, and toward completely new opportunities and aspirations, new concepts that might be joined to the ideas of equality and freedom in the minds and hearts of men.

Domestic quarrels, like those between labor and management, are a sad waste of useful energy, much like two dogs fighting over a bare bone in the middle of a chicken yard. Instead of fighting for a larger share of the current production, they might better cooperate to increase production as much as modern techniques and automation permit.

Quarrels between nations are much the same. They are stimulated by greed and fear, and they take the attention of both sides away from the obvious opportunities for cooperative progress. This is especially true of the U.S.A. and the U.S.S.R. Both have everything to lose and nothing to gain by failing to work together for peace and continued industrial growth.

Both have everything to lose—including living standards, leisure, security, education, and culture—and both may be vulnerable to threats and blackmail from the rest of the world. They may find themselves compelled to accept more responsibility for the welfare of the world than has ever fallen to one or two nations before, including the responsibility for maintaining world peace, and for helping other nations to achieve their own economic and political security.

Both have everything to gain by working together to solve the problems that face the world economy.

Although the two states started with entirely different economic theories and political philosophies, they are gradually finding out that the rules of the game are the same wherever it's played. In practice as opposed to theory, they find the same needs have to be satisfied and the same tools and procedures turn out to be most effective and efficient in both economies. As they develop and use scientific techniques for expanding their industry and producing more goods, for allocating and distributing the goods that are produced, and for borrowing the labor needed to maintain and expand their productive capacity, they will find it more and more difficult to avoid adjusting their theories to coincide with their practical experience.

In fact, their objectives are much the same, though in theory their priorities are entirely different. Both are striving to create strong economies that can provide a comfortable standard of living for all their people and that can satisfy the demands of other nations for economic assistance and cooperation. In theory, one is a planned economy and the other an economy of free trade, but in practice it is apparent that planning cannot eliminate the pressures of supply and demand, and that free trade cannot function efficiently in an automated industry without considerable planning.

By the end of the century it is likely that the two economies will be more alike than different in their actual operation though each may emphasize its own theoretically different point of view.

The two economies are like two similar machines with different ignition systems. The metaphoric spark plug in the U.S.A. is profit incentive, or the motivation of personal greed, whereas in the U.S.S.R. the spark is supplied by the motivations of collective greed and the need for collective power and security.

That profits are as essential to the economic growth of the U.S.S.R. as to that of the U.S.A. becomes apparent to anyone who gives the question serious thought. The difference between the economies has to do with how the profits are shared, and how the profit incentive is applied to stimulate the efforts that are needed for maximum growth.

The efforts needed include those of the planners, managers, skilled workers, manual laborers, teachers, researchers, scientists, and politicians, not to mention the philosophers, musicians, dancers, artists, and clowns. To move the economy forward, everyone must contribute his energy, talent, and enthusiasm, and the problem of each economy is to provide the optimum stimulation to every individual.

In addition to the material rewards implied in profits, it is possible to provide stimuli in the form of power, prestige, and honor, and these motivations are used lavishly in both societies. Status symbols are important everywhere, but living standards and wages are still the basic tools for getting a job done. So it is that in the U.S.S.R. as well as the U.S.A. there must be profits that can be distributed in the form of better living conditions and more comforts and conveniences and that can be invested in further growth.

How well the profits, the extra food, the new housing, the better roads, the education, entertainment, and research are distributed will determine to a large degree how successful each economy will be in reaching its goals. If the distribution appears to be unfair, there will be strikes, slowdowns, sabotage, and apathy in both economies; and to the extent that every worker feels that he is being rewarded fairly for his extra effort, there should be high enthusiasm and optimum growth.

When the political and industrial leaders of the U.S.A. and the U.S.S.R. recognize that their efforts need not be directed toward competitive goals, and that they can afford to cooperate in every possible way to increase their combined total production without any fear of running out of markets for their goods, then it is almost certain that they will find ways to increase the growth rates of both economies and to approach the levels that would be possible with the most efficient use of their combined resources of capital and labor. It is possible that some progress will be made in this direction within the next thirty-five years. If so, the chances are good that actual economic growth will exceed by a substantial amount the projections made in this study.

It is also likely that the lessons to be learned from the experience of the U.S.A. and the U.S.S.R. will be taken to heart by the rest of the world. Other nations will learn that poverty and deprivation can be eliminated in a modern industrial economy, whether it is in a capitalist or Communist country, and that benefits are obtained by making education available to everyone and by providing health insurance and the finest medical care for all of the people. Other societies will follow the example of the U.S.A. and the U.S.S.R., and will profit by their experience.

The objective of economic growth is to eliminate poverty and hunger, and to give everyone an equal opportunity to enjoy as productive and rewarding a life as his talents and abilities permit. Its objective is to give every man the chance to realize all the potentials of his mind, body, and spirit.

Its objective is only incidentally to produce more material goods, or to build great sports palaces, universities, and automated factories. All that a man needs is not yet known. He does need food, clothes, housing, education, medicine, and love. Until these are provided, no one will know what else he needs, and these are the first concerns of economies at all stages of growth that have yet been reached in the world.

Perhaps the time will come—sooner than we think—when it will not be possible to provide for further needs with more production and new factories. The need for integrity and self-respect is already more evident and more pressing each year, and the pace and direction of economic growth may have to be adjusted to these new requirements. Perhaps they are already being adjusted in small ways that haven't been recognized for what they are. The progress in communication, travel, and education is directed toward knowledge and understanding, as is the mechanization of industry and the exploration of outer space. If peace can be maintained until the end of this century, the economy of the United States will have reached a level of abundance from which additional production will be almost entirely for less tangible benefits than those that are its main concern in 1964.

If we can get to the end of the century, the prospects for the future that will be in view from that vantage point will be more breathtaking and inspiring than anyone dares to imagine now. In order to get there we will have to avoid nuclear war, and if we avoid that war, our economy will have done away with poverty and will have achieved a comfortable standard of living for all of the people.

In the future will be seen the reality of a society in which everyone can expect to get all the education he can accept, in which talents are discovered and encouraged equally regardless of race, color, creed, or status, and in which new measures of value are beginning to replace the financial and material measures that in 1964 are still characteristic of both the U.S.A. and the U.S.S.R.

Both economies are heading in a similar direction, and it is ironic that in this race, whichever reaches its goals of economic abundance and security first may find that the supposed advantages of being first no longer have any meaning. Unanticipated, new values will have been accepted by the world for measuring its leaders.

If there is peace these are the prospects for the U.S.A. and the U.S.S.R. The directions of growth and change are clear. The pace is not so certain. It may take only twenty years or it may take fifty to reach the levels projected for 1999, but in the perspective of history a difference of only fifteen years is insignificant.

The directions are what matter, and the consequent changes in opinions and prejudices and values that could amount to a new estimate of the potentials for mankind.

Glossary of Selected Terms

Brief Descriptions of Terms Frequently Used in Economic and Business Reports.

AUTOMATION—A general term used to describe a number of technical developments all of which substitute mechanical—mostly electrical—devices for human control of production processes.

CAPITAL—All economic goods existent at a particular time which yield realized or imputed income. Land, producers' goods (other than land), and consumers' goods. In a business sense, capital may refer to the investment in a business or to its net worth.

CAPITAL BUDGET—A budget limited to capital items. It separates capital or investment outlays and their sources of finance from expenditures and receipts used for current operations.

CONSTANT DOLLARS—A series is said to be expressed in "constant dollars" when the effect of changes in the purchasing power of the dollar has been removed. Usually the data are expressed in terms of dollars of some selected year or set of years. "Actual" or "real" dollars refer to prevailing prices.

CONSUMER CREDIT—Credit granted to individuals for use in purchasing consumers' goods and services; principally credit extended by commercial banks, retail merchants, cash-lending agencies, finance companies, and the service industries. The credit is granted for relatively short periods and is repaid either in installments or in one payment.

COST OF LIVING—Amount of expenditure required to pay for goods and services deemed to make up a given level of living, taking into account both changes in the quantities bought and the price paid. The term is often misleadingly used to describe a consumer price index which only measures price change, keeping the quantities constant over time.

DISTRIBUTION—Includes all of the activities involved in the passage of goods from the producer to the consumer.

ECONOMIC GROWTH—The sustained increase in the total and per capita output of a country as measured by its gross national product (in constant prices) or other output statistics. Rate of growth refers to the percentage of increase of one year over the preceding year.

FAMILY INCOME—Per capita income multiplied by the number of persons in the average family, estimated to be 3 persons in the U.S.S.R. and 3.4 persons in the U.S.A.

FULL EMPLOYMENT—A condition which exists when useful employment opportunities at prevailing wage rates are available for all persons who are able, willing, and seeking work. Since certain transitional changes cause some "frictional unemployment" at all times, full employment is said to exist when the number of job seekers does not exceed 3 to 4 percent of the labor force.

GROSS NATIONAL PRODUCT—Gross national product or expenditure (often abbreviated to GNP) represents the nation's total output of goods and services, valued at current market prices before deduction of depreciation charges and other allowances for business consumption of durable capital goods. The total reflects all goods and services produced during the current period, including a so-called "imputed" market value for goods and services consumed without passing through regular markets (such as the value of farm products consumed on farms, and the space-rental value of owner-occupied dwellings). Per capita GNP would be GNP divided by total population.

HOUSEHOLD CONSUMPTION—In this volume, a term applied to the U.S.S.R. to denote a situation in which consumer demand exceeds available supplies of consumer goods. The figures available for household consumption in the U.S.S.R. are roughly comparable to personal income figures in the U.S.A.

INCOME DISTRIBUTION—Refers to the distribution of families and individuals by income classes as well as to the distribution of total income by such classes.

INDUSTRIAL PRODUCTION—Refers to the physical volume of output in manufacturing, mining, and utilities industries. It excludes production on farms and in the construction, transportation, trade, and service industries.

INFLATION—An unstable condition in the general economy, characterized primarily by (persistent) increases in prices of commodities and services. This condition generally arises when increases in available money and credit (purchasing power are not accompanied by counterbalancing increases in the amount of goods and services offered for consumption.

LABOR FORCE—The civilian labor force (current Bureau of Labor Statistics definition) includes those of the civilian noninstitutional population fourteen years of age and over who are classified as employed or unemployed in accordance with BLS definitions.

MAN-HOURS, TOTAL—The sum of the hours worked by all employees. On a weekly basis, equivalent to average number of employees multiplied by average weekly hours.

MARKET RESEARCH—The collection and analysis of data relating to the distribution of goods and services, for the purpose of expanding sales volume.

NATIONAL INCOME—The aggregate return in money and in kind to labor and property for their contribution to current production; equal to the sum of compensation of employees, profits of corporate and unincorporated enterprises, net interest, and rental income of persons. Per capita income would be total income divided by population.

PERSONAL CONSUMPTION EXPENDITURES—As used in the national accounts, the market value of purchases of goods and services by individuals and nonprofit institutions and the value of food, clothing, housing, and financial services received by them as income in kind.

PERSONAL INCOME—Current income received by individuals, unincorporated businesses, and nonprofit institutions from all sources. Capital gains and losses are excluded. Personal income is the total of wage and salary receipts, other labor income, proprietors' and rental income, interest, dividends, and transfer payments.

PRIVATE DEBT (NET)—Gross private debt less duplicating corporate debt. Duplicating corporate debt is debt owed to other members of an affiliated system.

PUBLIC DEBT—*Gross:* Total government indebtedness including debt held within the government by government agencies, trust funds, and the like. *Net:* Gross government debt minus debt held within the government, by government agencies, trust funds, and the like.

REAL EARNINGS OR WAGES—The purchasing power of money earnings or the amount of goods and services that can be acquired with the money earnings; usually money earnings adjusted for changes in prices.

RESEARCH AND DEVELOPMENT—Basic and applied research in the sciences and engineering, and the design and development of prototypes and processes. Excludes routine product testing, market research, sales promotion, sales service, and other nontechnological activities or technical services.

> *Basic Research* includes original investigations for the advancement of scientific knowledge that do not have specific practical objectives. *Applied Research* includes investigations directed to the discovery of new scientific knowledge that have specific objectives with respect to products or processes.
> *Development* includes technical activities of a nonroutine nature concerned with translating research findings or other scientific knowledge into products or processes. Development does not include rou-

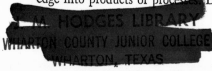

tine technical services or other activities, excluded from the above definition of research and development.

STANDARD OF LIVING—The kind and quantities of goods and services which are considered essential by an individual or group. Technically, standard of living differs from scale or plane of living, which is merely the list of things the individual or group consumes at a given time or place. But the term is frequently used in the latter meaning.

STOCK—Evidence of ownership in a corporation. Legally, it represents a contract stating the terms under which the corporation, as distinct from its owners, accepts capital from the stockholders.

TOTAL GOVERNMENT DEBT—The total of all formal debt obligations owed by federal, state, and local governments and, in the case of the United States Government, including the obligations of government corporations guaranteed by the federal government.